D1224401

Successful Smallholding

Planning, Starting and
Managing Your Enterprise

Successful
Smallholding

Planning, Starting and
Managing Your Enterprise

J.C. Jeremy Hobson and Phil Rant

THE CROWOOD PRESS

First published in 2009 by
The Crowood Press Ltd
Ramsbury, Marlborough
Wiltshire SN8 2HR

www.crowood.com

British Library Cataloguing-in-Publication Data
A catalogue record for this book is available from the British
Library.

ISBN 978 1 84797 075 6

Disclaimer
All tools and equipment used on a smallholding should be used in
strict accordance with both the current health and safety
regulations and the manufacturer's instructions. The author and
the publisher do not accept any responsibility in any manner
whatsoever for any error or omission, or any loss, damage, injury,
adverse outcome of any kind incurred as a result of the use of any of
the information contained in this book, or reliance upon it. If in
doubt about any aspect of any of the subjects covered in this book,
readers are advised to seek professional advice.

Line illustrations by Keith Field

Frontispiece courtesy Rupert Stephenson

Typeset by Jean Cussons Typesetting, Diss, Norfolk

Printed and bound in Malaysia by Times Offset (M) Sdn Bhd

Contents

Acknowledgements

The great pleasure in writing a book of this nature is the opportunity it gives you to revisit old friends and to make new ones – all in the interest of research. It also activates fond memories of personal smallholding experiences in the past and spurs one on towards new adventures in the future. No one can ever know everything, and it is only by accumulating information from as many different sources as possible that knowledge can be gained. Many enthusiastic farmers, gardeners, smallholders and country dwellers have assisted in the compilation of the book and the authors have cause to be grateful to them all.

Particular and effusive thanks go to Stuart and Julie Barker, David Bland (Southern Poultry Rearers), Carole Deedman, Val Porter, Steve Midgley and Bob Saxton. Thanks too to Pippa Barker of the Hampshire Beekeepers Association, Russell Fairchild, The Winchester & District Beekeepers' Association, Steve Irish and Neil Vigers for providing all the photographs related to beekeeping (Neil also kindly agreed to read the relevant chapter). Thank you to Adela Booth and the Jersey Cattle Society of the United Kingdom and also to Rupert Stephenson for once again allowing access to his not inconsiderable poultry photographic files. Roger Morris also conducted much valuable technical research on our behalf, for which we are, as ever, eternally grateful.

The several offices of DEFRA occasionally receive a bad press – but, from our own experiences, we cannot fault them and several necessary questions requiring a specific answer have always been dealt with efficiently and within the space of a single telephone call.

We are particularly grateful to Isabella Moreton-Smith and Stuart Webb both of whom have been kind enough to proof-read various drafts and to correct several elementary mistakes. Not content with that, they have offered several thoughts and observations all of which now appear in the text. Thank you.

Many of the above also provided unlimited access to their smallholdings; gardens and livestock in order that we might find suitable illustrative material. We are, however, also indebted to others who gave us photo opportunities that they will not be aware of. Several of the photographs shown here were taken at random while touring our home area of the western Loire in France, in Yorkshire and in southern England; if you think that you recognize your smallholding, allotment or livestock, then it probably is it and we thank you for providing such a perfect example of the point we were trying to illustrate!

We also much appreciate the efforts of Keith Field, our superb line drawing artist, who (goodness knows how) is able to turn our rough sketches into real works of art.

Introduction

Many of us have had 'the dream', the one that permits us to take our minds off a boring occupation and invades our lives, gradually and imperceptibly, until it becomes an obsession. The dream of starting a smallholding and managing our own little patch of self-sufficiency is one common to many, and this book is about helping you to achieve this dream. However, it is also about the possible pitfalls and realities of undertaking such a project and makes no attempt to view it through rose-tinted spectacles.

The smallholding trend began in earnest in the 1970s and owes much to John Seymour, the unchallenged guru of the movement, who wanted city people to venture into the countryside in order to enjoy what it had to offer. The trickle of converts desperate to drop out soon became a flood, but was, unfortunately, short-lived, due to their totally unrealistic expectations as to how much work would be involved. Although a smallholding may provide sufficiently for a retired couple, if your plans involve a young family and there is also a mortgage involved, you will certainly require an additional income from outside. It is, of course, possible for one partner to keep a well-paid city job while the other looks after the crops and livestock in the hope of, not only living off your smallholding, but also making sufficient money from the sale of surplus stock to provide an income.

It always pays to learn from the knowledge of others, and, between them, the authors have practised much of what they preach, as well as drawing on the experience of like-minded enthusiasts. In an effort to illustrate some of the potential setbacks awaiting those who hope to use their smallholding as an earner rather than as merely a means of creating a better and healthier life-style, the two following case histories may prove to be of interest.

A very good friend was an enthusiastic and knowledgeable gardener, always competing, in a friendly way, for the largest onion or the earliest tomatoes. He was, unfortunately, made redundant and so, encouraged by friends and family, decided to turn his hobby into a business. Indeed, he already had a regular clientele for his fruit and vegetables; he also sold eggs from his own hens, and so it was a natural progression therefore to step up into full-time employment. He rented a small paddock next to his house from a neighbouring farmer; invested in machinery, a garden tractor with implements, erected two large greenhouses and set to work. Sadly, he forgot to follow the first rule of successful smallholding and failed to establish his potential market. His local customers were insufficient in number to absorb the extra production and he struggled for two years to expand into new markets, while, in the meantime, much of his produce spoiled. Previously he had produced large crops of strawberries annually from a few dozen plants. He multiplied the number of plants by more than tenfold and, quite naturally, expected a tenfold increase in the fruit yield. The harvest was disappointing, and for the simplest of reasons: previously he had been able to lavish individual attention on each plant, giving each the best in the way of weeding, feeding and watering. Now his precious compost and his time had to be shared among a larger number of plants and there were just not enough hours in the day to give them the attention he previously had. The crops were still good, but smaller and the fruit not quite so attractive and he lost money.

In the second year he tried harder and decided to opt for the 'pick-your-own' market. His best fruit and most of his profits left the field in the stomachs of his customers. It would be good to report that the friend succeeded eventually, but, sadly, after four years he was forced to give up, after which he secured a part-time job in the local town. The lesson to be learnt from this story is not to run before you can walk, grow crops or keep livestock that require the least specialized attention, expand gradually, concentrate on good quality and spread your investment over as large an area as possible to lessen the risk of a single crop – you may not make a fortune, but you will not lose one either.

Customers are a law unto themselves: they can be your best friends one minute, but, if the produce in the next village appears to be a little cheaper, they will quickly desert you in droves. A second case history proves this point very well. The owner of a restaurant in a nearby town approached a successful, small-time, salad grower as he liked the quality of his lettuces, and especially the fact that he could pop along and cut them fresh as he required them. The system worked well, the grower planted a few extra each time and kept him well supplied without affecting sales to any of his regular customers. It was a long, hot summer and both grower and restaurateur did rather well. At the end of the season they met as usual to settle accounts and the restaurant owner told wondrous tales of expansion for the next year: more staff, more tables and an outside eating area. He invited the grower to supply even more lettuce; he would, he promised, 'take all you can grow'. The following spring arrived, but, sadly, the restaurateur did not. It was not until later in the season that it was discovered that the would-be buyer had negotiated a contract some miles away, and, because of the quantities he was demanding, could drive down the price, something the original grower could not and would not, entertain.

But it is unfair to dwell for too long on the negative aspects of the potentially 'commercial' smallholding. No matter what a particular year throws at you, never give up – if you have a bad crop, or a bad season, fight for your dream. In Rudyard Kipling's famous poem *If*, he says if you can 'watch the things you gave your life to, broken, / And stoop and build 'em up with worn-out tools' – lines very appropriate to the smallholder. The old farmers used to quote, 'one year in three you may make a shilling, the other years you just get by', but what they failed to explain is that in one year in three you may make money from milk and eggs, one year in three you may make money from sheep and one year in three you may make money from corn. Therefore, if you produced milk, sheep, cereals and poultry, then in most years it is possible to money from something.

If you work hard and are honest in your dealings you will get your reward, and it will not always be financial. Sometimes it will be something simple, like sitting in a hot bath on a stormy night after a hard day's struggle against the elements, knowing you have achieved something worthwhile. Sometimes it may come from sitting by the Rayburn in the early morning hours with a new-born piglet after being up all night while its mother gave birth to an otherwise fine litter; he is barely alive, but a weak pulse beats in his chest. You put him in a box of hay close to the warmth and go and get an hour or two's sleep. But when you wake, dress and come downstairs to look dubiously into the box and are greeted by a snuffling snout and a very pink, healthy looking piglet waiting for his breakfast, you can never put a price on that. Rolling back the tunnel cloches on a fine row of lettuce, all moist and shiny in the dawn light – even that is a good feeling and brings its own reward.

Unfortunately, enthusiasm and sentimentality alone are not going to be sufficient to guarantee a successful smallholding. You do need to be aware of the latest local and governmental changes and regulations. Attend smallholding courses and do not be afraid to seek the help of experts. Mix

'Home, sweet, home!'

tradition with technology – there is much on the internet, for example, that will inform and help on all of the examples above, as well as keeping you up-to-date with some of the latest media scare stories and hyperbole. Kept well and carefully, the livestock on the smallholding will not be as susceptible to disease and health problems as might be the case in a larger, commercial business. Nevertheless, it is as well know of some of the potential problems, especially if they may indirectly affect you by, for instance, the imposition of movement restrictions and legislation concerning the provision of disinfectant mats and the like at the entrance to your premises. The most obvious causes for

such activity are likely to be avian influenza (one of two notifiable poultry diseases, the other being Newcastle disease), foot and mouth disease (FMD) and 'blue tongue' – a problem transmitted by midges, but which cannot be passed from animal to animal and is no threat to humans.

Global forces are pushing the revival of 'hobby farming': you only have to look at the price charts of wheat, soy beans, cereals and dairy products. All of them show a flat trend line ending in a vertical rise of at least 100 per cent or more over the last eighteen months. High soft commodity prices are here to stay, even if, in the next year or so, we experience an economic slowdown. If ever there was a time to have a smallholding and grow your own, now is it. No wonder then that, in 2007, over a third of farmland was sold to what estate agents term 'life-style buyers'.

A final point which must be made is that this book can only be considered as an introduction to the subject of smallholdings and how to get the best from them. Each of the chapters really warrants a book of its own; in fact, there are many available in the Crowood catalogue that are dedicated to specific subjects in far more detail and it is recommended that you read at least one fully before you start.

How this book is organized
Having conducted several informal surveys regarding the particular interests of the members of a number of smallholding groups, clubs and associations, it appears that the majority are largely 'small' smallholders, with only a limited amount of space at their disposal – exactly the type, in fact, at whom this work is aimed.

There is a definite hierarchy of interests. Most are involved in growing organic produce and there is also as much enthusiasm for poultry-keeping, especially of the breeds of chicken that produce a regular supply of eggs

for the breakfast table. Bee-keeping is enjoying a resurgence in popularity and its followers are an eclectic mix of people. As to other forms of livestock, some smallholders have a goat, for example, and many more would like the opportunity to try their hand, but they are prevented from doing so by restrictions of space.

These factors have dictated the chapter layout. Vegetables, poultry and bees have been given their own because, as can be seen from the above, it is these subjects that will be the most likely options for anyone considering starting a smallholding, whereas the opportunity for the majority to take up the keeping of any of the animals brought together in the chapter on other livestock options will be limited. For those who are fortunate enough to be able to consider doing so, that particular chapter is divided into the several options, each of which is followed by a similar layout of sub-headings; this is deliberate and will, we hope, assist in weighing up the practicalities associated with a certain animal or breed.

J.C. Jeremy Hobson and Phil Rant
Summer 2008

CHAPTER 1

The Perfect Place

In an ideal world (which, as we grow older and more cynical, we all begin to realize does not exist), what would make the most suitable place for a smallholding? Would it be its size, location, topography, climate or a combination of all these things? In reality, smallholding at any level will more than likely be a matter of compromise.

Be realistic regarding the best ways of improving what is already there – it may be necessary to improve the drainage before considering any enterprise, and you need to be aware of the possibility of frost pockets and other such weather vagaries. There is, for example, little to be gained by the planting of an orchard at the base of a slope, which is, or so you fondly believe, protected from frost by a wall of a building on its down side – planting an orchard on a slope is good, but planting at its base in such a situation is not. Spring frosts literally drain downhill, as the air just above the ground becomes very cold and slips between warmer layers above it. Far from offering protection, the wall will allow frost and cold air to build up against it, to the detriment of any young trees not yet above its height. And that is just a simple example.

There is also the possibility that you may wish to keep livestock. Fencing for a few hens is not likely to be a problem, but it could be a different matter when considering the keeping of a couple of goats, which are renowned escapologists. Stock fencing, unfortunately, does not come cheaply and so, when searching for your perfect place, it makes sense to look at properties which have had the necessary work done on them, rather than to think blithely that it is a job you could undertake later when you have a spare few minutes.

LOCATION

Location is important for an array of reasons. You may be in the fortunate position of retiring or selling up in the more financially rewarding parts of the country and downsizing to a rural retreat. You may, for the sheer practicalities of having to remain where you are for work and the children's education, need to continue living in your present home and consider, instead, the prospect of renting an allotment or re-evaluating the potential of your existing garden. Whatever the situation, some factors remain the same.

A south- or south-west-facing location is obviously the ideal, no matter whether you are considering smallholding on a Scottish croft or vegetable growing in an inner city garden. A north-easterly situation will not encourage animals, bees, fruit or vegetables to thrive. Does the area you are contemplating flood on occasions? If it does you can bet that it will exactly at the time your strawberries are beginning to form or your young livestock just so needs that small patch of young, inviting, fresh grass.

Relocation

Relocating to a place in the country and finding a cut-price smallholding has become the

Goats are renowned escapologists and will require some secure stock fencing to keep them where they should be.

dream upon which many modern-day 'reality' television programmes are made. The would-be relocators draw up a list of essentials and then ask the presenters (in truth, their researchers) to find them the ideal situation – a tall order, bearing in mind the fact that they have not, in all probability, thought things through. Is, for instance, relocating realistic, even after one has considered all the demands of the several family members and the almost certain need to earn money? Can we do the physical work and are we really sure about this 'journey into the unknown'?

Assuming that all these questions (and more) have been answered in the affirmative, you next need to consider in what direction your own particular Utopia lies. Mountains and moorland are, for example, more suited for grazing hardy livestock than they are for growing delicate and fancy fruit. A smallholding in such a location is not likely to be close to a bus route or in an area to which teenage children feel drawn. In addition, isolated farmsteads are not conducive to the easy and effective marketing of produce.

A Renovation Project?

If you are thinking about the purchase of an old property with a view to doing it up, will its location allow access for any heavy vehicles that may be required to bring in machinery or materials? When considering the possibility of buying and renovating a truly remote property that has no mains water at present, bear in mind the fact that a connection to the mains may prove expensive. The actual cost depends on factors such as how far the property is from the mains and whether the pipe will have to run through someone else's land. As with any kind of outside contractual labour, it pays to get a definite quotation before agreeing to any work. Money has an unfortunate habit of disappearing once any kind of renovation work is being considered, and so always allow a larger budget for any particular project than one might at first assume was needed.

If you are in a position to contemplate buying a property to turn into a smallholding, where will you live until the improvements have been made if it is currently uninhabitable? Is its location close enough to your existing dwelling for you to be able to travel to-and-fro while either doing the work yourself or keeping an eye on professionals doing it?

It is always worth having a word with any long-established locals in the area you are looking at in order to find out from them whether the local planning authorities are sympathetic towards building changes. Not only is it necessary to keep on the right side of the Department for the Environment, Food and Rural Affairs (DEFRA) (as will be seen towards the end of this chapter), but it is also important to keep abreast of any local politics – so start by establishing friendly relations with members of your parish council.

An isolated smallholding may not be ideal for a young family or for those with produce to market.

It is not always possible to run a smallholding in such an ideal situation, with a couple of hectares, there is plenty of space for livestock and produce.

SIZE MATTERS

By now you might be forgiven for thinking that this book is, despite assurances to the contrary on both the jacket notes and in the introduction, aimed solely at those fortunate enough to seek out and purchase a life-changing property. It is not, and the smaller the amount of land you have at your disposal, the more that size matters.

With several hectares, it matters little if a piece is taken there for a vegetable patch and a site allocated there for a poultry run and, while we are about it, why don't we earmark that rough piece for hard standing and a tractor shed? Unfortunately, the average person does not have these advantages and, quite

literally, every single square metre is of great importance. In such a situation, crop rotation is vital; the positioning of a chicken run away from where it is likely to upset the neighbours is crucial, and a friendly farmer on whose land you might be able to temporarily locate a hive of bees is essential. If that same farming friend is ready and willing to rent out a piece of ground on which to rear some sheep then all well and good, but have you considered the fact that you will probably require a trailer in which to transport them? If so, in the interest of security, the best place to keep such a vehicle is most likely at home next to the family car. Is there room? The perfect place for you is obviously where you live, close to work and schools, but it would

WATER SOURCES

A picturesque well or bubbling spring may offer a private, and more importantly, free source of water. It is unlikely that, in this day and age, any but the remotest and most dilapidated of properties will be without mains water, but a free natural supply will obviously cut down on costs when it comes to irrigating the vegetable patch and providing water for animal troughs. No matter what its intended use, it is advisable to check the quality of a spring or well before you buy a property and also to establish how reliable the well or spring is – both can run dry in a drought or if the underground water course changes direction. Also, underground water sources can alter as a result of building works, climate change or flooding.

There are many ways a natural source can be 'harnessed': a pump is the simplest answer in the case of a well as the water can be taken

not be seen as being so perfect by the seeker of a smallholding with a bigger agenda. What suits one will not suit all, and so, despite whatever smallholding dreams one might harbour, it will be necessary to cut your coat according to the cloth.

Water is essential and can be collected from a number of sources in order to water stock or vegetables. Here it is being collected from the roof and stored until required; a tap is set at such a height as to allow the use of a bucket or watering can.

direct or stored in a header tank. If a spring is on your land, then the water obtained from it is, in the same way as that from a well, your property. Provided that one is not contravening any local by-laws or those imposed by the water authorities, it may even be possible to divert water from a stream, and we know of several smallholdings where this has been done in order to make it feasible to install a waterwheel and generate power. For the majority, however, that may just be a step too far.

Annual checks are a good idea, especially if it is intended that a natural water source be used for animal drinking water or is to be piped for use in outbuildings or utility rooms. As well as the health aspect, high levels of limestone, for example, can damage heaters, boilers and washing machines. Water checks can be done professionally, but it is also possible to do your own. If you would rather test it yourself, home water analysis kits give a quick result and can test for nitrates, copper, iron, chlorides, pesticides, acidity (pH level) and hardness. The kits include litmus papers, which change colour when dipped in water and can be read against a chart showing recommended levels. The quickest results show within 10sec but others, such as the bacteria test, can take up to 48hr. If you are worried about the test results from a private water source, contact the help line shown on the test packet.

Even on land that does not contain an existing spring or well it may be possible to have a borehole drilled. To find out if and where there is water under your land you can consult the local water authority and ask whether they have any maps of underground watercourses. A more entertaining option is to employ the services of a water diviner: some we have witnessed in action have subsequently been proved to be absolutely spot-on, while others have been a little less reliable. It is worthwhile mentioning that, generally, any water found which is less than 5m (16ft) under the surface is unlikely to be drinkable; it should be between 5 and 20m (16 and 65ft)

underground. However, you cannot test the water quality until a borehole has been drilled, by which time things will have already become expensive.

SAFETY AND SECURITY

Over the years, we have had so many encounters and conversations with fellow smallholders that we now believe that perhaps the most important criteria (and ones that have never, as far as we are aware, been seen or heard included on anyone's 'wish-list') are those of security and safety. All livestock enthusiasts have experienced or know of a near accident involving farm animals, and many have, unfortunately, experienced an actual one, as a result of a beast's escaping from the smallholding and reaching the public highway. As may be imagined, a frightened cow, donkey, sheep or goat is not likely to have much road sense when it stumbles upon rush hour traffic, and the situation can quite easily have a fatal outcome, leading to at the least the loss of a valuable and favourite animal and at the worst, legal prosecution. In our opinion, all smallholdings on which it is intended to keep livestock should be surrounded by fenced grazing and situated up a private drive, both ends of which are blocked by securely fastened gates. Of course, only in the rarest of circumstances is this situation ever likely to be reached, but it is a basis from which compromises may be made.

Seclusion, while it does have obvious advantages in respect of making potential highway disasters most unlikely, can, however, lead to problems of security. A sequestered property is a magnet for not only thieves wishing to steal tools and machinery from a barn or outbuilding, but also – and thankfully rarely, yet nevertheless devastating for those to whom it has happened – for those few perverted and disturbed members of society who, for some reason, gain some strange pleasure from the killing or mutilation of animals and poultry.

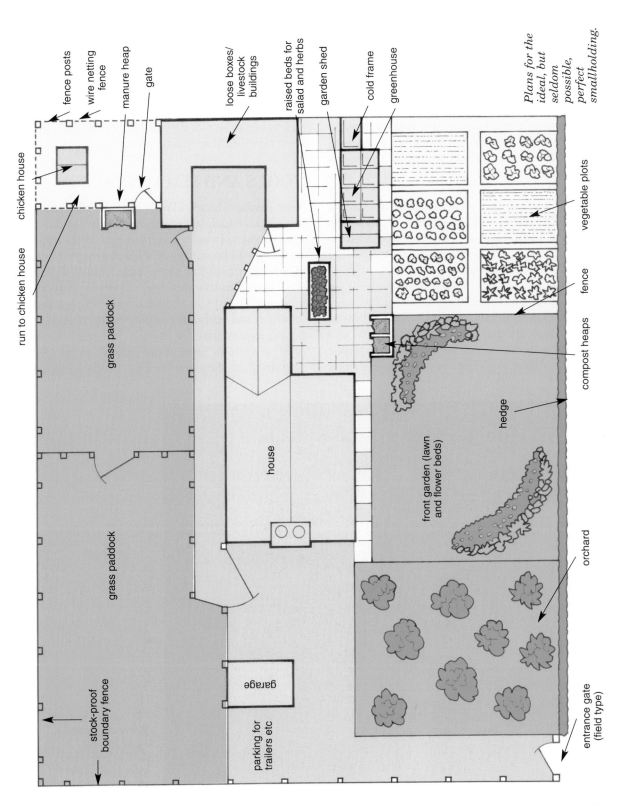

fence posts

wire netting fence

manure heap

gate

loose boxes/ livestock buildings

raised beds for salad and herbs

garden shed

cold frame

greenhouse

Plans for the ideal, but seldom possible, perfect smallholding.

chicken house

vegetable plots

run to chicken house

fence

grass paddock

compost heaps

hedge

front garden (lawn and flower beds)

house

orchard

grass paddock

garage

stock-proof boundary fence

parking for trailers etc

entrance gate (field type)

17

Footpaths and Bridleways

Land over which the public are allowed access will not, by definition, be all that secure. Although the average country walker will respect livestock and close all gates, there are undoubtedly some who will not. It pays to consult an Ordinance Survey map in order to establish what rights of way you may have over your own couple of precious hectares, especially if it happens to be a Byway Open to All Traffic (BOAT). A BOAT allows walking, horse riding and the use of wheeled vehicles of all kinds, a situation which is not amusing if the route is used on occasion by local, off-road motor-cycling groups or 4 × 4 enthusiasts.

There are other categories that allow access over your land, among them ordinary footpaths and bridleways, restricted byways, green lanes and permissive paths. Should you ever, at any time, need to plough up a public path or in any way cause it to be inaccessible, you must reinstate it as soon as is practicable. Although the landowner is not generally liable for any accidents caused by a potentially dangerous path, he is if they caused it to be dangerous, so make sure that the reinstatement is done in an effective manner if you wish to avoid the possibility of an accident claim falling through your letterbox.

TOOLS AND MACHINERY

Your perfect place will obviously require some tools with which to keep the soil tilled and the grass mown. In addition, it might be thought necessary to be equipped with chainsaws and other labour-saving tools. Your smallholding will need suitable storage (of which more in Chapter 7), but you should be considering the logistics and economics of buying or of merely hiring equipment for which you may have only spasmodic use. A neighbour might be prepared to carry out some cultivation or

There are several right of way definitions that allow access over your land.

Useful Tools

Rotary tiller: if you have a small amount of land to cultivate, this is perfect for the job, but ensure that it is powerful enough for the work you have in mind; do not even consider one with less than a 5hp engine.

Backpack sprayer: a back or knapsack sprayer is an extremely useful item and has several general uses, such as watering, weed-killing (be sure to clean it well after use), fertilizing and disinfecting livestock sheds.

Brush-cutter or strimmer: you will almost certainly need a powerful, petrol-driven machine to keep weeds and undergrowth at bay.

Chainsaw: it goes without saying that a chainsaw is a dangerous piece of equipment; if you have any serious cutting to do it is advisable to call in an expert, otherwise, read the safety and operating instructions and practise at ground level until you feel confident.

other tasks, such as hay-making, which will save you the expense of buying equipment.

The normal tools-of-the-trade, such as spades, shovels, forks and rakes, are essentials and you should buy the best you can afford. Consider stainless-steel garden forks and spades, they are a joy to use, easy to keep clean and, although initially expensive, an extremely good investment. At the very least, buy a spade with a forged steel blade, as cheaper spades have a blade made from pressed sheet, which will often buckle and bend under strain. Likewise, an ordinary garden wheelbarrow is unsuitable for the smallholder, not only is it too small, but it probably will not withstand the weight it will be asked to carry. A builder's barrow is probably the best option.

Tractors

Do you need a mini-tractor, or even a 'proper' job? You might if your smallholding is more than a couple of hectares, but, other than that, one could be seen to be an indulgence, a 'big boy's toy', the money for which could more

If you have a large area to work, it may be worth considering the purchase of a small, second-hand tractor.

profitably be used elsewhere. If you do have a large area to work, it might, however, be worth considering a small, second-hand agricultural tractor (a new one would be astronomically expensive). An old tractor may bear the scars of a working life and not be as powerful as a newer model, but its advantages are many. It will be more powerful and probably less expensive than a mini-tractor and require only a little tender loving care before it is capable of some hard work. You will also be able to use it with the implements common to all tractors, which can often also be bought quite cheaply from local, second-hand dealers or sales and auctions, or you may be able to borrow them from a neighbour in return for some help when he is busy. The only disadvantage in buying second-hand stock is that there is unlikely to be any warranty, but, on the other hand, most of the depreciation will already have occurred and, should you ever wish to sell it, you will lose very little – in fact, with the growing popularity of small tractors such as the Massey-Ferguson, for example, you may even make a profit.

Safety Courses

The agricultural and smallholding world is awash with safety courses, and quite rightly so. Power tools are dangerous things and should be treated with respect and their workings fully understood. The easy option is to look on the internet for a suitable course near you or to contact your local agricultural college and enquire what one-day or weekend courses they do. Quite often these courses, which may concern subjects such as tractor-handling or working with chainsaws are Lantra-certified and all comply with the requirements laid down by the Health and Safety Executive. (Lantra is the Government-licensed Sector Skills Council for environmental and land-based industries.)

SMALLHOLDING COURSES

Not to be confused with the safety courses

mentioned above, these offer an insight into the world of smallholding or, as self-sufficiency is becoming increasingly known, self-reliance, and are invaluable in teaching specialized country skills. They give a true taste of what smallholding is all about and cover topics as diverse as basic plumbing, livestock husbandry, hedge-laying, bee-keeping and growing organic crops. Having at least a basic understanding of these subjects (and there are many more) will undoubtedly give you invaluable knowledge and confidence, and will, perhaps more importantly, save costly mistakes at a later date.

These courses vary in duration, some which are privately run can be a week in length, but most seem to be planned over a weekend period. Like the safety courses, many are held at agricultural colleges and organized by local farm training groups. Look out for their being advertised in periodicals such as *Country Smallholder and Smallholder*. It is a good idea to attend as many introductory courses as your finances will allow, not only will you learn from them, but it is an extremely good way of getting to know like-minded people, any one of whom might prove to be a useful ally and kindly giver of advice in the future. Some of the courses will also help to explain the legislation behind the registering of your smallholding and obtaining flock numbers, which can appear daunting and complicated to the newcomer.

REGISTERING YOUR HOLDING

If you are intending not to restrict your smallholding activities simply to vegetable growing, bee-keeping or poultry husbandry, but to attempt any sort of livestock management, it matters not how small your smallholding is going to be nor whether you intend keeping one animal or a dozen, you will, in principle, still need to apply to the Rural Payments Agency (RPA, a government department working within the auspices of DEFRA) in order to obtain an agricultural holding or CPH (county parish holding)

Smallholding courses cover many related topics, including ditching. Here 'before' and 'after' pictures show what can be achieved.

number. This is a nine-digit figure; the first two of which relate to the county, the following three the parish, the final four being a unique identification of you.

Telephone DEFRA to obtain the telephone number of the appropriate county office who will be responsible for allocating your holding number. Having made contact, you will then be sent a form on which to apply, together with a pre-paid envelope in which to return it once completed – it may be seven to fourteen days before one is issued, but while you are waiting, console yourself with the knowledge that registration costs nothing.

It is better to get into the habit of referring to registering an 'agricultural holding number' rather than a 'smallholding number' as, technically, there is no such thing as a registered smallholding. Even those wanting to keep sheep, pigs and goats in an urban or semi-urban area should still, when making the initial enquiries, use the term 'agricultural holding'.

The issuing of a holding number is essential for administrative purposes and makes

No matter whether you have three or thirty head of livestock, you will require an agricultural holding number.

the whereabouts of any animal much easier to trace in the event of a disease outbreak. Should you have cause to contact DEFRA, the use of your CPH number should, all things being equal, make it easier for them to access your details. Having obtained your holding number, it is necessary to contact the AHO (Animal Health Office) who will issue you with a herd/flock number, relevant to your holding. This is used to trace both the keepers of animals and their individual animals; it is possible to be issued with the same number(s) for cattle and sheep, should you be intending to keep the two, but pig herd numbers follow a slightly different pattern.

Pets or Livestock?

Technically, you can keep pets but not farm animals without registering them. DEFRA considers a goat, for example, as a farm

animal and is thus not a pet; so, if you own just one you should, in theory, obtain an agricultural holding number and a flock number. There are a great number of people who do not comply, but in a book of this nature, it is right that the authors should insist that you (dear, law-abiding reader) stay well within the law.

Having obtained both a holding and a flock number, you will then be required to keep records of the movements on and off the holding, the drugs administered, drug batch numbers, dosages and administration dates, withdrawal periods and be subject to inspections from your local trading standards office (in reality, the AHO). All animals must be ear tagged and, if the European Community eventually has its way, they will need to be multiply tagged with different tamper-proof tags. Already, new double-tagging rules have been introduced as of January 2008, which, alongside a change in the requirements for movement licences and record keeping, promise to ease the burden on farmers and smallholders by making the forms easier to complete. (For more on specific ear-tagging procedures see Chapter 6, p.87).

ENVIRONMENTAL WASTE

To add to all this, there are now new Environmental Waste Regulations applicable to a holding. Once you have a holding number the Environment Agency will request you to seek permission to spread manure on your land, store animal and plant waste on your holding or a host of other things that farmers have been doing from time immemorial. As further examples, if you wanted to burn your hedge trimmings, fence posts (not tantalized ones) or tree branches, spread the waste from the ditches which you have cleared or put rubble down on farm tracks or field entrances which you have brought or bought in from outside

your holding, you would also have to apply for exemptions.

Should you consider applying for rural payments on your newly acquired holding you will also be required to have your land inspected by DEFRA (the AHO division) before they will allow you to keep any farm animals. It may well be that they will do this even if you are not applying for any grants. On the grant side, it will also be necessary to be aware of the Nitrate Regulations which include stocking density, animal droppings and water courses.

The Single Payments Scheme

Most of the RPA activities are involved with the administration of the Single Payments Scheme which is a voluntary programme designed to pay farmers for managing the land in an environmentally friendly manner. It is doubtful whether it would be worthwhile for a smallholder to register for this scheme, despite the lower limit being just 0.3ha (0.7 acres). Also, as mentioned above, compliance

Ear-tags are required on all livestock, including goats.

with the scheme would dictate the width of field margins and a host of other technicalities that get in the way of managing your farm or smallholding. Fines levied by DEFRA for non-compliance can be heavy and, in the case of an under capitalized smallholder, probably fatal.

ROSE-TINTED GLASSES AND REALITY

After reading what constitutes the ideal and perfect smallholding at the beginning of this chapter and the potential difficulties involved in, and because of, possibly needing to obtain an agricultural holding and flock number at its end, the reader might be tempted into throwing up his hands in horror and living his smallholding dream through television programmes such as Hugh Fearnley-Whittingstall's 'River Cottage' series or the BBC's 'It's Not Easy to Be Green'.

It is certainly not our intention to scare anyone off (for one thing, it would negate the need for this book), but merely to make readers aware that, by pursuing certain avenues, one might encounter previously unthought of legislation. It is important to realize therefore, that, as with most things in life, it is all a matter of compromise when it comes to finding and working the ideal smallholding. In reality, it will be few who can fulfil their dreams in their entirety, either through lack of capital, acreage, time or commitment. Almost all can, however, achieve some degree of smallholding satisfaction in the smallest of places through vegetable gardening, fruit growing, poultry husbandry and even bee-keeping, all subjects covered in detail throughout the following chapters.

Even the smallest garden will provide a certain degree of self-sufficiency.

CHAPTER 2

The All-Year-Round
Vegetable Patch

Some form of vegetable gardening is possible on the smallholding or even in the back garden. An allotment is the perfect place for a vegetable plot, but the availability of them is limited in some areas in the country; this is not surprising when recent figures tell us that 200,000 allotments have been lost in favour of building over the last thirty years. There is, apparently, a ten-year waiting list in some places. The good news is, however, that, under the 1908 Smallholdings Act, if six individuals make representations to the local council, it is legally obliged to make provision.

Assuming therefore that there is land of some description available somewhere in the vicinity, the main aim of the vegetable-growing smallholder must be to extend his season in order to yield early and late produce, which, if you are hoping to sell some of your surplus, attracts a higher premium and therefore, one hopes, a greater profit. It is also obviously important to make the most of your growing space, maximize your cropping and make the task as easy as possible.

ORGANIC GARDENING

There is much hyperbole in the media concerning organic produce, and possibly much confusion in the mind of the consumer about what it actually is. In its basic form, organic gardening is a method of producing food without recourse to synthetic chemicals such as those found in fertilizers and pest-control sprays. By choosing not to use them,

one attempts to encourage the existing soil bacteria and natural predators to do the job for you. Not only is this kinder to nature and the environment, it must also be kinder on our bodies. To ingest quantities of chemicals with our food inevitably causes health problems, some of which are probably as yet not apparent. But, above all, the authors believe that the taste of organic food is better. The final benefit is that it is expensive to use such chemicals, whereas natural methods require a much smaller financial outlay.

While the main objective behind the organic system is to build up and maintain good health in the soil, there may be occasions when this is not enough. It may be, for example, that a crop of brassica plants seems to be at a standstill at some point in their growth and needs a little extra boost. Under the chemical system it is no problem – open a bag of granular fertilizer (in this case perhaps high nitrogen), sprinkle a little around the plants and the plants recover. But such treatment is also possible under the organic regime as there are many proprietary organic fertilizers available. Many organic gardening books will give recipes for the creation of liquid fertilizers, ranging from animal manure to nettle leaves and more, steeped in rainwater buts. The solution is later diluted and watered around the plants, and many do indeed work as well as, if not better than,a chemical alternative. The heart of the organic garden is the compost heap or, more accurately, heaps. Properly constructed and

25

ABOVE: Organic food growing is better for the environment and for you, too much use of chemicals in the vegetable garden could eventually, perhaps, see you on the wrong side of this dividing wall.

LEFT: The 'heart' of any garden, but especially an organic one, is undoubtedly the compost heap. Without a regular input of humus, which the heap provides, there will be a decreasing amount of goodness left in the soil with each crop that is harvested.

tended, the compost system can produce many of the nutrients required for the growing of an organic garden.

Identifying 'Good' and 'Bad' Soil

The first objective of anyone intending to grow vegetables, whether they be organic or not, must be to improve the productivity of the soil, you will be very fortunate indeed if you have purchased or are in possession of land that is perfect for your needs. It may be stony or waterlogged, it may be exhausted by years of poor management, or simply a 'thin', dry and 'hungry' soil, on which even weeds fail to thrive. Much can be perceived by walking the land in a growing season: if the soil is rich in weed growth (in particular, nettles and docks), it may be in good heart but in need of a thorough cleaning before planting.

Evidence of poor drainage will be seen by the presence of rushes and members of the buttercup family (*Ranunculaceae*). Dry, thin soil will have a sparse covering of weed and grasses, and a trial dig with a spade will reveal a poor surface structure and an underlying, light subsoil. All land can be improved and the first task is to ensure that you do not waste valuable time, effort and money on improving soil that you do not intend to cultivate. Mark out your plan for the area on the ground before you begin. One of the best layouts for production, particularly if the plan is to concentrate on hand cultivation, is the 'deep bed' system.

The Deep Bed System

This is based on cultivated beds, normally about 1.2m (4ft) wide and as long as you wish, separated by permanent pathways. The width is important because of the need to cultivate the beds from both sides without your having to step on to the soil. Planting can be carried out by using a board or boards laid across the bed to avoid compacting the soil, particularly in wet weather. A useful tip when deciding upon the exact width is to make it according

Make the beds just wide enough to fit the width multiples of your cloches.

to the width or length of any cloches you may be intending to use. And by making the beds uniform, you will soon know the exact number of plants needed for each row. Because there is no need to leave access spaces between the rows, more plants can be grown per square metre than would normally be the case.

By far the greatest advantage of this system is that it makes the correcting of deficiencies so much easier; for example, one can remove stones from stony soil and use them to improve the permanent pathways. Poor draining soil can be improved by raising the level of the beds above the pathways. The sides can be supported, if required, by boards, baulks of timber such as old railway sleepers or even concrete blocks. Hungry or exhausted soil can be improved by the addition of large amounts of humus, in the form of compost or well rotted animal manure. As to the healthy but weed-infested land, dig out as many as possible of the roots of perennial weeds such as nettles and docks, cover the levelled surfaced with a thick layer of newspaper or even the cardboard from boxes laid flat to suppress the weeds. Then cover this with a layer of clean topsoil or compost and plant into that. The paper/cardboard will eventually rot down, having sufficiently suppressed the weeds beneath long enough for them to die.

Raised Beds

Similar to the deep bed system, raised beds can be useful in small areas and will grow salad crops and herbs very successfully. Because of the nature of their construction, they can be built near the kitchen door, saving trips to the other end of the garden in driving rain. Raised beds can be adapted to suit any size of available space and have the added advantage of looking attractive, thus forming a feature in the garden rather than detracting from it. Railway sleepers make the ideal border, being strong, robust and heavily preserved, of the right length and, cut in half, of almost the exact width (if you are ever tempted into cutting one with your chainsaw,

use an old chain as the preservatives, the hardness of the wood and the odd piece or two of gravel that has attached itself to the tar, will ruin a new one and no amount of sharpening will get it back into condition). Now ever-increasingly difficult to get hold of, in the absence of sleepers, you might be forced into constructing a bed with breezeblocks or some alternative material.

If starting from scratch, make them of any length, but, as with the deep bed system described above, never more than 1.2m (4ft) wide; the whole point of a raised bed is that the soil should never be compressed by your standing on it and so it is necessary to be able to reach the middle from either side. If you are building a bed on concrete, an existing yard or a patio perhaps, the height needs to be about 60cm (2ft), but, otherwise, less than half that depth will suffice. With some existing soil at the base, it will take less to fill with good quality and expensive compost. In future years, the beds can be kept topped up by the addition of home-produced material from your composing bin, but, to begin with, it will probably be necessary to buy sacks of the stuff.

EXTENDING THE GROWING SEASON

Having produced the ideal environment for growth, it is now necessary to examine the ways of extending the growing season. The lack of light will always slow down growth at the end of the year and it may be financially prohibitive to attempt to alter that. Commercial growers, with their economies of scale, can afford the outlay on glasshouses or poly-tunnels, but, on a small scale, it would be difficult. There are natural ways of improving one's chances, and, if you have selected your site with care – preferably facing south, you will capture all the available light in the autumn and spring periods. Ensure your site is open, that is, not overshadowed by trees or high hedges. A little artificial help will, however, pay dividends. Cloches, for example,

A cloche or garden frame will extend the growing season of lettuce.

will extend the growing season for such plants as early lettuce, where, with even the cheapest type, one can produce a crop two to three weeks earlier than in the open ground, thus making the ultimate goal of self-sufficiency more of a reality.

Cloches

One of the most effective ways of extending the growing season is by the use of cloches. These fall into three main categories: first there are the sheet or film type consisting of a length of plastic sheet and a number of wire hoops. The ends of the hoops are pushed into the soil on either side of the row of plants, the plastic sheet covers the hoops and is tied off to a peg at both ends before a second lot of wires is placed over the sheet and hooked into loops at ground level – much in the same fashion as the base of a Boy Scout's tent. To keep the wind from getting under the thin plastic sheet it is advisable to partly bury the bottom of the sheet, all along the length of each side. This type of cloche is used in horticulture on a large scale where the whole arrangement is put in place by special machines. The advantage is low cost, the disadvantage is possible wind damage and the fact that the sheeting can normally be used for only one crop or one season – the plastic soon becomes dirty and damaged and is therefore replaced each year.

The second type, not seen so often nowadays, but common three or four decades ago, is the glass 'tent' or 'barn' cloche. Special wire frames are used to hold two sheets of glass like a tent or four sheets like a taller barn. The frames have a handle to permit the cloches to be lifted off the crop. Their greatest advantage is that they can be periodically removed for weeding, feeding or replanting. With planning, they can be moved from the row to be harvested on to the next one needing protection and they are fairly resistant to all but gale-force winds. The disadvantages are that they can easily be broken through careless handling and that glass is no longer cheap. In frosty weather especially, they can become damaged if the glass is frozen to the ground.

Some General Benefits of Cloches

Cloches also permit you to keep the crop cleaner; free from mud splashed up by the rain and prevent wind damage to the foliage. As a result, the crop also looks better and will be more attractive to customers. Many crops can be given an early start under cloches, which are then moved on to help the next successive planting. Crops, such as peas, carrots and the cabbage varieties, all benefit from the early warmth and protection of cloches and will, as a result, get away to market sooner. In the summer cloches are invaluable for cropping strawberries and they also protect your valuable fruit from bird damage. Early outdoor tomatoes can be started under cloches until they touch the cloche, while, at the other end of the season, ripening fruit will be assisted by laying the plants on a bed of straw and covering them with cloches until more fruit has ripened.

The third type is the rigid, plastic cloche; some are also made from translucent fibre-glass and are very strong. They have many advantages, just a few of which are the facts that they are easily movable; if they are kept clean, will last for many seasons; are difficult to break and very effective in use. Being lighter than glass they may need to be held down in windy weather and are usually supplied with pegs for this purpose. Their main disadvantage is cost, but as this can be spread over many seasons this may be acceptable.

Garden Frames

The next step up from cloches is the garden frame – a sort of miniature greenhouse. These can be home-made and many an old window frame has done service as one. The sides can be constructed of thick timber – old floor boards are ideal – or one can construct a more solid frame from brick or cement block.

The Victorian kitchen garden could not have survived without a plentiful supply of garden frames.

Ensure that you are able to prop open the 'lights' (the glass panels on the top) on fine days, as ventilation is important in order to deter mildews and other rots of this nature. The frame may be cold or heated by the use of heater cables either buried in the soil or pinned around the inside of the frame close to the soil.

The advantage of frames is they can be used to protect tender plants from frost and also to raise early plants in the spring, in much the same way as a greenhouse. Once the frosts are over, the frames can then be planted up with such items as melons, which rarely do well without extra protection. The soil in garden frames should be dug out once a year and replaced with fresh, compost-rich soil to avoid the build up of disease and to replenish the nutrients. If you have access to fresh animal manure, a good thick layer in the frame topped up with clean soil will result in some heating of the frame as the manure rots down; done at the right time of year (early spring), this heat can assist the raising of young plants in pots or trays on the frame soil. By the time the seedlings have gone out to permanent stations, the manure will have rotted down and will feed the permanent frame crops for the summer.

Greenhouses

The most versatile and important building any gardening smallholder can construct is the greenhouse and it has the advantage over frames in that it can be comfortably entered, and it is therefore possible to enjoy its shelter and relative warmth while carrying out any necessary work. With the addition of some form of heating and lighting, it is possible to not only lengthen the growing season but also the length of the normal working day. Greenhouses are usually constructed of aluminium or timber and glazed with either glass or plastic. Timber frames tend to be warmer as they do not conduct the inside heat out though the frame, but they may need more maintenance. They will also fractionally reduce the available light due to the fact that wooden frames tend to be thicker in their construction than greenhouses made from aluminium. Aluminium has the advantage of being almost maintenance-free and some manufacturers now fit their products with exterior plastic 'bar capping', which holds the glazing in place and reduces the heat loss through the glazing bars. Glass glazing needs only an annual wash to ensure the maximum light transmission, but beware of too much light availability as it could scorch your plants on a hot day. You may need to consider shading during the summer months. Plastic glazing does not transmit so much light, but can trap more heat.

Poly-Tunnels

Smaller versions of the poly-tunnels used by commercial growers may have their place on the smallholding. Rather like very large cloches that one can walk into, they lend themselves to the raising of crops grown in the soil inside and are superb for raising winter salad crops. As is the case with traditional greenhouses, they can be fitted with heat and light and will, as a result, permit one to work out of the weather, as well as extending both the working day and the growing season. They are cheaper to purchase than rigid framed greenhouses, but the covers will require replacement about every four or five years. Beware of installing them in very exposed gardens, as they are easily damaged by storms, and perhaps with a loss of your precious crops.

IRRIGATION

Many good gardening books mention that soil should be free draining but moisture retentive. This appears at first reading to be a confusing statement and may therefore benefit from some explaining: the object is to keep a constant film of moisture around the roots of the plant, but, obviously, very few plants thrive in soil that is waterlogged; in addition, they need oxygen around the roots. Imagine a section of topsoil suspended above the ground

Asparagus needs free-draining soil and the rows are usually ridged in the spring before the first shoots appear. Grow it only if you have the space and patience; you will have to wait at least two years for the first cropping.

in a sieve, water the top of the soil until the water comes out of the bottom. Now let it drain until no more water comes through – this soil is now at what is known as 'field strength'. Allowing for evaporation, theoretically, if one added one drop of water to the top, a drop of water would exit through the bottom. This is the state we wish to maintain.

To achieve this happy condition is not, unfortunately, always so easy and it may well be necessary to improve the water-retention qualities of the soil. Bear in mind the fact that, generally, sandy soils are poor water retainers whereas clay soils are very good, but can have a tendency to become waterlogged. The qualities of both can be improved by the addition of humus; that almost magical element found in compost, well rotted

FYM (farmyard manure) and peat or peat substitutes. The best soils contain a mixture of differently sized grains, clay soils have very small grains and tend to stick together, or to flocculate, making them difficult to keep open to water and air. Sandy soils have larger grains which tend to let water straight though. Humus mixed into the soil will separate the fine grains in clays and will fill in the spaces in sandy soils, making them better at retaining water.

Watering

Despite the high rainfall in the United Kingdom it is rarely enough to keep the soil at 'field strength' without assistance. There are many irrigation systems on the market, from simple sprinklers to large, oscillating nozzles that fire large amounts of water over a large area. It is important to consider the needs of your system before investing. Many have a very coarse droplet size and this can not only batter down young plants, but will also pan the soil surface, causing it to become an impervious layer from which the subsequent water will just run off. The watering should be gentle and thorough and the newly applied water should soak down to the bottom of the root ball of the plants. Once irrigated, avoid adding further water until the top layer has dried; during this time the ever active smallholder should be out with his hoe breaking up the surface pan, destroying weed seedlings and reducing the moisture loss from the soil beneath by creating a 'dust mulch'.

The practice of giving everything a light sprinkling every evening is a waste of water as it rarely, if ever, permeates down to the root system before evaporating. Also, if the watering is constantly in the top few centimetres of soil, the roots will be forever struggling upwards in search of moisture rather than forcing their way downwards, making for weak and unproductive plants. The nearer they are to the surface, the more prone they are to drought.

By far the most efficient, and effective, use of precious water is the 'seep hose'. This may be either a special hose made to weep water along its length or a flat hose with small slits cut along it. Either way, the hose is laid along the row of plants and the water seeps into the soil exactly where it is required: no water is wasted watering pathways or lawn edges. The initial cost may be high, but, with care, it will be repaid many times over.

Holding Ponds

As can be seen throughout *Successful Smallholding*, your source of water is important, and, even in the smallest of vegetable patches situated within a few metres of the family home, you may not always be able to rely on mains water, and a hosepipe ban during the growing season could spell disaster. The

The Importance of Hoeing

The one tool which is essential in your vegetable gardening armoury is the hoe. Its frequent use will ensure that weeds are disturbed long before they have a chance to become a nuisance, and that the top inch of soil around your valuable crop will form a dust mulch, which, in turn, reduces water loss through transpiration – hence the old gardener's saying, 'A good hoeing is worth a shower of rain'. If kept clean and sharp, the hoe is a joy to use, and an hour spent each morning working between the crops ensures that by the afternoon the tiny weed seedlings you have disturbed are 'frizzled' by the sun and eventually return to feed the soil.

larger smallholding may experience even more difficulties, and, even if you have a stream or river through your property, you may not be permitted to draw water from it during times of drought. Water butts and bins may not hold a sufficient reserve to tide you over a dry patch.

If it is possible during the planning stages, you should explore the possibility of storing water in a reservoir sited near to the house and buildings. A simple, evacuated pond, with the sides built up with the spoil could serve well. If you then route all the rainwater downspouts from the house and buildings the winter rains could supply all you needs during a dry spell. There is no need for expensive liners for the pond or reservoir, many companies sell reinforced plastic tarpaulins which they will supply to size. A submersible electric pump fitted with a filter on the outlet should give sufficient pressure and volume for your needs, but check the capacity with the supplier before purchase. (One word of warning: if you are using seep hoses attached to a home-installed reservoir system you should consider fitting a filter to the main water pump since sediment in the water can block the pores.)

LEFT: Water butts may not be sufficient for a season's supply. Any water stored will be kept fresher if light is minimized; black polythene goes some way towards achieving this.

BELOW: Collecting rainwater from any available source will, however, help during the drier months. The lengths of wood seen in these water barrels are not there by accident, but placed deliberately in order to give amphibians, small animals and birds the chance to escape after an unexpected ducking.

Mulching

Although not technically to do with irrigation, mulching does, nevertheless, provide a practical way of ensuring moisture retention in the soil and is a way of making more use of the water available. A mulch can consist of a layer of almost any organic material, compost, well rotted FYM or lawn clippings placed around the crop. It helps to suppress weeds and, by shading the soil, will reduce water loss by evaporation. Worms from beneath will gradually draw the mulch into the soil thus helping to improve its fertility. The only drawback may be due to the fact that it can also act as a good daytime hiding place for troublesome pests such as slugs and snails. At the end of the season, any loose material left from a season's mulching can be removed to the compost heap; but if you have a poultry run, throw it to them first, they will soon remove any pests lurking therein.

COMPOSTING

Commercially constructed composting bins vary somewhat in effectiveness and tend to be quite expensive; nevertheless, the better ones should last a lifetime and you may as a result consider them to be a worthwhile investment. However, with just simple, basic carpentry skills you will be able to fashion your own and the materials required may be obtained for little cost if you search around. Depending upon the size of your holding, you will require at least two bins which should be sited in a convenient corner, as you will be visiting them often. Practicality and space usage dictate that they should not occupy the best piece of your valuable growing area and a shaded area under trees is fine. If the area is low and wet, it is worth taking the trouble to incorporate rubble and stones into the soil before construction in order to aid drainage.

Anything that has once grown can be used to make compost, the secret is a good mixture of green and fibrous material to avoid wet lumps, and the turning of a compost heap from time to time will speed up the process of decomposition. Ideally, the mixture should be moist without ever becoming sodden – a simple cover of old lino, carpet or plastic sacks will keep out the worst of the rain and permit

In order to provide a regular supply of compost for your garden at least two bins will be required.

Be Careful What's Included in the Compost Heap

It is worth noting that, to conform to certain EU/DEFRA regulations, if you have free-range poultry which might have access to a compost heap, the heap should be totally fenced off in order to prevent the birds from picking at offal such as meat scraps, as the feeding – accidental or otherwise – of animal to animal is illegal.

the material underneath to generate the heat necessary to turn straw into compost gold. If you plan to keep livestock on your small-holding then their waste products, such as bedding and excrement, will be a valuable addition to the project. Poultry manure in particular is a very good compost activator, but it should not be used until it is well-rotted.

CROPPING FOR A MARKET

With the large variety of seeds and plants available today, it is possible to produce many crops virtually all the year round. This is particularly true of the brassica family, where it is far better to grow, say, a constant succession of cauliflowers to satisfy a regular local market than it is to grow a single, huge crop once a year. In this way you spread your risk and have a more regular income. The one crop, 'cut and clear' system is more suitable for the larger grower who may be supplying a supermarket chain.

Salad Leaves and Tomatoes

With the growing awareness of the value of fresh produce and a healthier eating regime, a constant supply of salad leaves of all varieties is vital. No longer does the lettuce rule supreme; even this noble salad staple has undergone a change – the choice available in both hearted and loose leaved varieties is enormous and the customer is becoming more adventurous and discerning. This is the market you should pursue. With protected cultivation one can produce lettuce all year round, along with the very fashionable rocket, land cress and many other salad leaves. Aim for the maximum variety you can effectively produce then concentrate on those that give the best return in your area. Once you are known for them locally, your crops will virtually sell themselves.

The humble tomato has also attained new status. Commercial tomato crops are grown with specific requirements in mind. A large crop is the foremost one, the tomatoes must be even in size and shape, and should store well, that is, have a long shelf life and the skin must be tough enough to withstand bruising. Low on the list of priorities will be taste. You, on the other hand, are in a different market and can grow for flavour. There are many novelty tomatoes which have grown in popularity – striped ones, yellow and orange ones, even a dark, almost black one. They increase the visual interest of a salad as well as having a fine range of flavours and will add value at the point of sale. While you may be unable to supply tomatoes all year round, you can certainly extend the cropping season with protected cultivation. The greenhouse is an ideal example, but even outdoors crops can be laid down in the rows on a bed of straw and covered with cloches to permit them to ripen at the end of the season. Also bear in mind that even green tomatoes have a market for the chutney makers in your area.

Potatoes and Root Crops

New potatoes always receive a premium over main crop potatoes: with the use of cloches or protective fleece you can sow new potato seed early, followed a month or so later by an un-protected crop and then, later still (up until the end of June), plant a further batch which will produce new potatoes in late September.

Root crops can also have their season extended. By early sowings under cloches you can be the first in your area selling young carrots and, at the other end of the season, still be selling them, once again protected

Whatever summer crops you decide to grow, try and ensure that there is a regular supply of them rather than a one-off glut.

under cloches. It is all about getting your timings right and being quick to grab the opportunity of a dry day early in the season when most people have not given their garden a thought.

WINTER

Eventually, the long, cold nights will slow down or stop all growth; however, this may be one of your busiest times. The temptation to sit by the fire and peruse next year's seed catalogues may be overwhelming, especially if accompanied by a drop of Scotch or glass of wine, but this is the time for preparing for the new season. Compost and FYM dug in now will have a chance to be incorporated into the soil over the winter months and the turning of the soil will break up heavy, cloddy lumps – an action that will be further helped by a few heavy frosts. With deep beds there is no need

to walk on and compact the soil, working from each side just lightly fork your precious compost into the top inch or so and leave all level and clean.

Greenhouses and frames can have their soil removed and replaced with fresh, once again, mixed with your own compost or FYM from the dung heap. The newly added soil will then have plenty of time to settle and warm up. At the same time, give your greenhouse and garden frames a thorough cleaning and remove any old shading before sterilizing them with a proprietary, organically-friendly product such as Jeyes Fluid. Clear out every nook and cranny that may harbour pests and diseases. Likewise, clean and sterilize all unused flower pots and seed trays and put them to dry and air well before storing them.

Winter is the best time for bonfires. Burn those woody cuttings that were unsuitable for the compost, together with the cabbage stumps that you had thoughtfully hung up in the poultry run, now dried and too woody for compost. Remember when the ashes of the bonfire are cooled, put them through a garden riddle and store them in the dry, they are an excellent source of potash and will be relished by soft fruits and potatoes later in the new year. Nothing should be wasted, the aim is to export as much as possible and to import as little as necessary into your venture.

Check your tools and clean them thoroughly; if you need to make any new purchases, choose them well and be very selective in what you purchase. Beware of cheaper, pressed steel spades and forks – they may be fine for the occasional weekend gardener, but are unlikely to stand up to the rigours of regular hard work. Look for forged steel, or better, stainless steel if your purse will run to it. Handles should be hickory or ash; some cheaper items have softwood handles which may give under the strain. Beware also of tool systems which feature a handle with a locking device on to which many tool heads can be fitted, these may look attractive, but personal experience has shown

that they are really suitable only for light use. An investment in good, strong tools will repay itself many times over. If they are cleaned and oiled after use and kept stored in the dry they will last a lifetime.

Finally, perhaps you can be permitted your spell by the fireside since winter is, indeed, the time for planning, for ordering seeds and for judging which crop made a good profit and which ones did less well. Compare your planting dates and, having made any adjustments for next season, mark the new calendar to remind you – a fortnight lost can never be regained. By taking time to read about some of the old gardening practices, such as complementary planting, the knowledge gained could help in preventing pest and disease problems in the future: planting marigolds next to onions, for example, is said to prevent infestations of onion fly, and it has been scientifically proved that their presence kills off certain harmful nematodes found in the soil. Maybe on Christmas Day you can permit yourself a holiday, raise a glass to your success and try to learn from your failures. But remember the new season starts tomorrow!

The older gardeners made much of the fact that complementary planting will help to prevent pest problems. Placing marigolds next to onions is perhaps the best known of these.

CHAPTER 3

Creating an Orchard and Growing Fruit

The delights and virtues of an orchard are extolled briefly in Chapter 7 and there is no doubt that the practical and visual aspects of a smallholding are greatly enhanced where one is incorporated. It does, however, require some regular maintenance and, even when grazed, the grass will probably need topping at least once a year and the fruit trees will certainly need some skilled pruning. If you leave your trees to grow unchecked there is a real danger that storms will break some of the branches, which will result in damage to the trunk of the main tree. Aside from that factor, however, pruning is required to ensure the formation of young wood on which the blossom and, eventually, fruit will form.

It is unlikely that the local authorities who let out allotments will permit the growing of fruit trees on their plots, as they normally have a policy of not permitting anything to be grown which cannot be taken away easily, at the lease-end or when requested to do so. Elsewhere, all kinds of fruit growing is possible – even in the smallest of gardens. There are numerous varieties of fruit trees that can be grown and trained against a house wall or left permanently in tubs. In fact, growing trees in tubs could increase your options, as it may be possible to consider some of the more delicate types if you have a porch or sheltered corner into which they can be wheeled for protection in the winter. When choosing varieties suitable for your own situation, remember to buy either two or choose a self-pollinating type.

Soft fruits are easy to grow in tubs, buckets or, indeed, in any similar type of container and take up little space. A couple of plants in each receptacle will provide a daily picking throughout the harvest season and, if different types are chosen, it is possible to create an extended cropping period. Any pot-grown plants will obviously require more regular watering than those planted direct into the ground, but as they are being grown on a small scale, an automated system of watering could be a simple matter to arrange. The soil will also need regular feeding as the growing plants will quickly leach it of any goodness.

PLANNING FOR FRUIT TREES

When planning for fruit trees, look for a fairly open site well away from the roots of other trees and shrubs that could compete for the soil's nutrients. While it will pay to cut back any grass or weed growth from the proposed site immediately before planting (perhaps even resorting to spraying with a herbicide such as Roundup in order to do so), unless it is absolutely essential, do not plough or otherwise cultivate the plot. Any soil cultivation other than that immediately where trees are to be planted is required only where the soil structure is poor or compacted.

Alternatively, most fruit trees will grow well when trained against a wall of the house or outbuilding, provided that their roots are firmly fixed in a good bed of soil. The heat reflected back from the wall will help to ripen

Beyond the scope of most, but a walled garden is the perfect place for training espalier fruit trees, or even, as here, the inclusion of a glass peach house.

the fruits. Apples and pears grown either as a hedge of cordons or as a fan or espalier against a wall will take up little space; but to restrict the amount of pruning you need to do, it is important that you choose a tree grafted on to the right rootstock Some fruit trees are self-fertilizing, others require the purchase of male and female plants – although you might get away with a single tree if your neighbours have similar ones in their garden as the local bee population will transfer the necessary pollen from one to the other. Most container-grown apple trees are sold ready trained because it is the first two to three years' growth that requires the most skill in cultivation. By the time you buy them, all you really

need to do is to decide upon the eventual shape. The size depends on the type of root-stock to which they have been grafted. You will find that there is less choice with pears since most are grafted on to a semi-dwarfing quince rootstock.

The financial outlay will quite often dictate the size of the trees being considered, although specific situations may call for some forward thinking well beyond the scope of economics. On windy sites, for example, it will pay to plant smaller trees which tend to establish better than do larger specimens – they also have a stronger root system that is less likely to be disturbed in windy conditions. Otherwise, if the wallet can afford it, plant container-grown trees of around 2m (6ft 6in) in height. If space allows and you are creating a 'traditional' orchard, trees in such a situation are normally planted at the rate of rather less than 150 per hectare (equivalent to about 360 per acre), which works out at a distance between trees of around 9m (10yd).

Planting

Having got your fruit trees home, dig a hole for each of at least three times the size that the plant roots currently seem to demand and remove all stones from the excavated soil (a coarse riddle will help). Some would advise minimizing any soil disturbance at all and planting through the clipped turf of previously sprayed grass, but we believe that, at

the very least, you should break up the soil and remove any large stones from it before returning it back around the tree roots. If the roots of your new plant have encircled the pot, tease them out gently, making sure that the root ball is moist before planting. Remember to break up the soil at the bottom of the hole as well as that which has been removed; if the soil is particularly poor add a handful of bonemeal or some slow-release fertilizer, and, as you fill up the hole with the new soil mix, firm it in around the roots and base of the plant as you do so (a further application of fertilizer will benefit the tree at a later time, particularly during its second season). Each tree should be planted at a slightly lower than the level it was in its

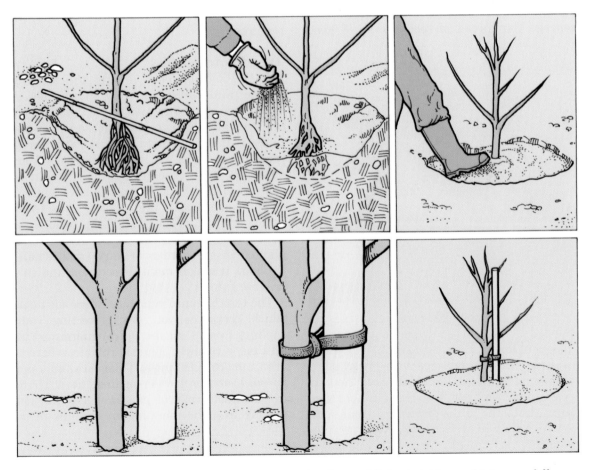

Break up the soil and remove any large stones before planting and (below) staking and tying carefully.

nursery pot, and, if the top growth is so luxuriant that it is likely to act as a sail in the wind and thereby continually disturb the roots as they struggle to establish, thin out some of the branches in order to reduce the wind resistance (but do not cut out the central leader or you will end up with a very strangely shaped tree).

Staking

Generally, fruit trees will need to be staked so that wind movement does not cause the roots to become unstable in the soil. However, this is not always the case, as on a windy site, for example. A larger tree (sometimes known as a 'standard') should almost always be staked, but not with a tall post that will totally prevent the tree stem from swaying in the breeze. Some movement is necessary to encourage the spread of root growth. A good

analogy would perhaps be to suggest how, on a windy day, you are sometimes forced into standing with your feet apart in order to counteract the force of the gale. Use short stakes protruding no more than 30–45cm (12–18in) above ground. Take care with the ties used to secure the stem to the stake: the wrong type or one tied incorrectly will chafe the bark and do permanent damage to the tree. Specially designed ties are available from garden centres and agricultural suppliers, although even these will need to be adjusted from time to time. Strips of hessian or webbing are good alternatives and can be attached to a stake with a clout-head nail to prevent them from slipping up and down both the post and sapling in windy weather. Do not leave the trees staked for too long either, one full growing season should be all that is required.

Tree Protection

A rabbit-netting fence around the orchard will certainly protect it from the unwanted attentions of rabbits and hares, and, if you are ever intending to graze geese within its confines, it will prevent them from wandering off (but not foxes and wandering dogs from entering). It will also fail to protect the trees from deer and is a very expensive option. Where some kind of protection is considered necessary (and, of course, it will probably not be a consideration for the mini-orchard planted in the garden or very small smallholding), it makes more sense to use individual tree guards.

The type of guard you choose depends upon which creatures you are protecting your saplings from. A plastic, spiral guard may be all that is required against rabbits, whereas even a more substantial Tuley tube may not be sufficient protection against deer. There are available rolls of plastic tree-protection netting that the manufacturer claims to be

To protect any trees against damage from livestock some kind of tree guard must be incorporated.

biodegradable; it is, but not for at least seven or eight years – by which time it will have embedded itself in the tree if positioned too closely in the first place. As a consequence, do not set any type of wire guard so close to the trees that new foliage becomes entangled in the mesh or, as just mentioned, the wire becomes embedded in the bark. For protection from grazing stock there is no alternative but to surround the tree with a guard made from three or four tantalized stakes and a similar number of rails or battens, a length of stock fencing or sheep netting and some barbed wire – the last may, one hopes, discourage the worst excesses of rubbing, an activity in which all forms of livestock seem to delight.

But once a tree guard is complete you cannot just walk away and forget about it. It needs regular maintenance. Weeds should be pulled out, supporting stakes checked and any dead or diseased branches removed. Keep an area of at least 1m (3ft) in radius surrounding each tree weed-free for two to three years. A carefully handled knapsack sprayer and suitable herbicide might be the most practicable means of doing this, but for personal reasons you may prefer to use a mulch of organic material. Once the trees are in production allow grass to grow as it reduces the (unwanted) uptake of nitrogen by the tree. Additionally, grass actually conserves the soil's fertility and provides a habitat for potentially useful predator insects.

Pruning Mature Trees

Fruit trees, if you intend to maintain them and get good crops, need careful attention. In addition to the winter pruning, when

In a mature orchard grass can be allowed to grow right up to the tree trunks, provided that it is kept closely cropped.

Without correct pruning, long, thin branches will be unable to support the weight in a good fruiting year, causing some limbs to break or become misshapen.

diseased and crossing branches are removed, summer pruning is equally vital. This regular trimming, especially for plums and cherries, is straightforward enough and is done when the sap is rising so that cuts heal quickly, thereby reducing the risk of fungal spores entering the wound. The object is to make the tree produce short, fruiting spurs and not carrying out this work will eventually lead to long, thin branches laden down with fruit and bending down towards the ground. Shortly after congratulating yourself on the magnificent crop, you will find half the branches torn off with its weight and the tree spoiled for years until you can reshape it.

Read the pruning section in your gardening book before you start for specific, detailed instructions. Generally, however, and with the use of only good quality tools, it will be necessary to take the offending branch back to a joint and cut it away neatly. If you intend to cut wood thicker than your little finger invest in some good loppers, preferably of the Bypass type. For wood thicker than this, buy a good pruning saw – do not try to force loppers through thick material as this is likely to rip the bark and damage the cutting tool. Once the limb has been safely parted from the tree, if you have any rough bits remaining, clean them off smoothly. To leave open 'snag' wounds from the teeth of your saw creates ideal opportunities for disease to become established. The practice of painting wounds went out of fashion many years ago as modern research has proved that this actually seals in disease (although some older books still advocate the practice). Instead, leave the cut clean and, in time, the cambium or bark layer will close over the wound. Fan-trained plums or cherries against a wall simply require their side-shoots to be tied back against canes or horizontal wires fixed to the wall. In the winter all you will need to do is to shorten the main stems and branches by roughly one-third of the current year's growth to encourage new side shoots.

SOFT FRUITS AND FRUIT BUSHES

It would be a shame if you did not take advantage of the summer months to grow some soft fruit such as red and blackcurrants, raspberries, blackberries or gooseberries. A fruit cage in which you can grow all of these without the risk of their being stolen by the birds could be made to fit into any odd corner. Ideally, it should be at least 5m × 5m (5.5yd × 5.5yd), but some thought should be given to its planning.

Planning

Most soft fruits enjoy a fair mount of sunshine so it is important that the taller bushes do not shade the smaller ones. The more sunshine you get on your fruit, the riper it becomes and the sweeter it tastes, and so, depending on the positioning of your garden, start with strawberries at the front of the plot (facing south, if possible) and lead up to the highest (probably raspberry canes) at the northern end. There are many ready-made, sectional fruit cages available, the least unobtrusive and longest lasting being those made from tubular steel. Some come with permanent netting fixed to the side panels, others require a net to be placed over the framework.

Choose your varieties carefully: with some types of soft fruit, such as blackcurrants, there are dwarf varieties that are excellent for the small garden, but, when purchasing

Fruiting spurs on mature apple and pear trees are thinned out during winter to prevent the possibility of too heavy a fruit crop.

for the smallholding, it is possible to buy varieties that will grow up to 1.5m (5ft). Gooseberries are now being grown as 'thornless' – a prospect which makes their harvesting that much more attractive. Both can be grown as fans, in the manner described earlier in this chapter when we considered fruit trees, which again, makes them ideal contenders for the small garden, or, in fact, anywhere where space is limited. Some soft fruit selection may depend on your topography and weather conditions, as spells of severely cold weather are essential to stimulate the flower spurs of some plants into action (blackcurrants are, perhaps, the best known of these). The older varieties of gooseberry may be badly affected by mildew, which often reduces the next crop to nothing, so you can help yourself by going for a disease-resistant variety. Plant breeders are constantly developing varieties that are disease-resistant so it is always worth asking your supplier for the most up-to-date advice.

Planting

In an ideal world, you would begin by planting bare-rooted plants at the end of February or the beginning of March, but with the easy

Watering

Do not forget the potential difficulties of watering; although perhaps not as essential as in the vegetable garden, walking great distances carrying water or unravelling unsightly, tangled hosepipe in order to water your newly planted shrubs and trees will soon lose its novelty.

availability of container-grown, soft fruit bushes, it is, provided that they are carefully watered in a hot summer and possibly protected from the worst of the weather during their first winter, possible to start your fruit cage at any time. Beds for soft fruit must be well dug and have had lots of well rotted manure or compost added to them. This will have the effect of raising the surface level a little higher than your path, but this is not a problem since the enriched soil can be kept in place with either edging boards or by being regularly raked back into gently sloping

sides. Black polythene, buried at the edges and through which slits can be cut to accommodate bushes and other fruit-bearing plants, will ensure that there is little weeding required, virtually no mulching and only the minimum of watering.

Staking

Most cane fruits need supporting rather than actual staking, and some a combination of the two. Summer raspberries, for example, need the addition of some sort of climbing frame and the easiest way is to hammer in posts about 2m (6ft 6in) in height and 2 to 3m (6ft 6in to 9ft 9in) apart before stretching thick gardening wire horizontally between them at equidistant heights. The canes are then tied and trained along these wires as they grow. Autumn raspberries, on the other hand, need no permanent supports and can be treated more like an herbaceous perennial. But they may in a particularly abundant season, require to be held up by means of strategically placed stakes and a loop or two of strong garden string.

Depending on their type, raspberries may need some support when fruiting.

Here used in the vegetable patch, plastic strips may help in protecting soft fruit from the unwanted attentions of small birds.

Protection

Birds love soft fruit, and unless this is grown in proper wire cages it may be necessary to net it by using fine-mesh nylon net on stakes. Black cotton strung backwards and forwards between the fruit bushes, both high and low, will deter birds who need to fly into it only once to be frightened. Other ways of protecting crops from the unwanted attention of birds are by the use of bags on sticks and of old compact discs left to rotate on strings so that the silver coating catches the light and startles would-be avian predators as they fly over.

Home-made scarecrows are effective provided that they are regularly moved around – one of our best successes was a realistic-looking scarecrow accompanied by his 'dog', made from a 5gall plastic drum, four pieces of roofing batten, a large plant pot and an old furry rug. It is sometimes possible to buy plastic kestrels from your agricultural suppliers (you can also find plastic owls to stand guard on posts around vulnerable areas of the garden). By poking two tall and whippy hazel sticks into the ground and suspending one of these birds on thick black cotton stretched between them, the slightest wind will move the sticks and the kestrel, making it seem as if it was swooping and hovering over the ripening crop. A neighbour here in France places a stuffed fox in his fruit bushes at the appropriate time of the year, but that may be going a little too far.

Pruning

The pruning of soft fruit varies greatly with the type being grown. However, for most fruit pruning is light for the first year, then in the following years from a third to a half of the length of the main shoots is removed. Your gardening book should have a chapter on pruning and will advise on specific pruning times and methods, all of which are dependent upon the type and age of the fruit to be pruned. Gooseberry bushes over three years of age, for example, should be pruned in the winter, and the best way of doing this is to cut out one in three of the thickest looking stems at the base of the shrub. Red- and blackcurrants will benefit from a little winter pruning, but only about 8cm (3in) should be removed from the tips of the main stems. Prune well established autumn-fruiting raspberries in February by cutting the old canes right down to the ground and then tying in the new stems at waist level in order to support the next crop. Prune blackcurrants in October, once the leaves have turned colour, by removing a portion of the older stems right down to the ground, or low on the bush at a joint where a strong side shoot is emerging. As with most garden tasks, a little forethought will lead to good results; neglect only leads to tears and disappointment.

PESTS AND DISEASES

In addition to the general garden pests and diseases mentioned elsewhere here, some fruits have their own particular problems. A general spraying programme is often advised, but many common fruit pests are difficult to control by these methods as they arrive on the wing and disappear just as soon as they have laid their eggs. One old, traditional way of trying to cut down on the numbers of insects, aphids and disease-carrying spores was to build a fire in the orchard or fruit garden so that the smoke blew through the trees and caused any grubs present to roll up and fall off. The old gardener's tip of a winter tar oil spray may help and will certainly kill off any dormant eggs. It should be done in December or January.

If you do want to try and cut down the risks of potential disease in a slightly more scientific way than that, then it would not hurt to apply a fenitrothion-based spray (but not on cherries when they are flowering) or gamma BHC dust. Fenitrothion is a contact insecticide and is a member of the organophosphate family. It is both non-systemic and non-persistent, is compatible with other insecticides and is available as a dust, an oil-based, liquid spray or in a soluble, powder formula. Gamma BHC dust is extremely useful on vegetables and fruit trees, eradicating wireworm, weevils, chafers and leather-jackets. Mancozeb, applied in the spring, undoubtedly helps in preventing leaf blotch in cherries. This is one of the non-systemic (surface-acting) fungicides, known as ethylenebis-dithiocarbamates – thankfully known as EBDCs for short. Their use will control many fungal problems, including blight, leaf spot, rust, downy mildew and scab.

Organic Alternatives

You need to decide at the very outset whether you will use sprays or rely on more organic methods for pest control. If you choose the latter, then your results may not be so effective, but at least your conscience will be clear. Improve your chances by selecting tree and shrub varieties which are known to be the more disease-resistant, and remember that good management will produce better stock more able to resist attacks of any kind. In addition, encourage the natural parasites of potential pests (ladybirds, for instance, are known to feed off greenfly – even so, no pest predator can be effective in controlling a pest until it has built up its numbers), pick off any diseased leaves or remove caterpillars by hand and burn dead material so that it does not become a breeding ground for unwanted fungal or parasitic visitors.

Many garden centres and agricultural suppliers now stock pheromone traps that use the smell of the females of the species, such as codling moths, in order to attract the males. Codling moths are particularly troublesome on apple trees where they lay their eggs on the immature fruit. These eventually turn into small maggots that burrow into the apple as it ripens. The trap works in the same way as the old, traditional flypaper and the male, attracted by the smell with which the trap is impregnated, is caught on sheets of sticky material. The males having been caught, the females are obviously unable to mate and, in turn, are unable to lay eggs. One trap, which is left hanging throughout the summer months, for every four or five trees is usually enough. They are also effective against other, similar species of pest and well worth the small investment of a few pounds. Grease bands placed around the trunks of apple trees will also trap the grubs of winter moths as they attempt to hibernate in the ground at the base of the tree in the autumn.

STORING FRUIT

Some fruits will store for months when kept in the right conditions – and they need not be that elaborate (in Chapter 7 it is suggested that 'a garden shed is all that is required. It needs a strong concrete base and, if it is intended to use a part of it as a place for storing harvested crops over winter, it must be situated somewhere where there is the least likelihood of frost'). It therefore needs to be cool, with a constant temperature; ideally, around 3–7°C (37–45°F) and certainly nowhere where the temperature is likely to

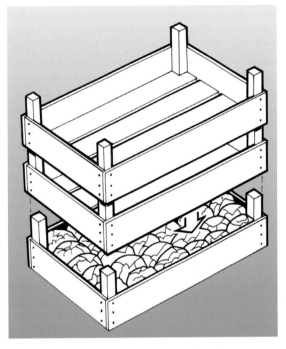

Ensure that there is a circulation of air between each box and also that the fruit is stored in single layers rather than piled on top of one another.

What to Store

The apple is the best, closely followed by pears. There are many late ripening varieties of both that will keep for a long time – some for as long as a year. The time of harvest is, however, all-important; the best 'keepers' are not actually ready to be eaten when picked; in fact, were you to harvest fully ripened apples they would not keep so long due to their starches and acids having been already allowed to ferment into sugars, whereas, when under-ripe fruit is stored, it slowly ripens over a period of months.

---

fall to below 2.8°C (36°F). Wherever is the most suitable, be it a garden shed, garage, unused loose-box, garage or cellar, should be well ventilated, dark and with a moist, but not damp, atmosphere. If necessary, damp down the floor in order to prevent shrivelling; alternatively, store the likes of pears and apples in polythene bags (having first ensured that they were harvested on a dry, rather than humid, day). If you are using polythene bags, a constant temperature is even more important: pierce one hole per 500g (1lb) of fruit and store in a single layer so as to avoid possible condensation.

How to Store

Never try to keep anything but the best – one bad apple can, quite literally, spoil the rest. Choose unblemished specimens and throughout their storage check them regularly and be prepared to remove any rotten or diseased individuals. The secret is to store everything on a single level in either boxes made for the job or the strong cardboard boxes in which fruit has already travelled to your supermarket or local greengrocer. All of these boxes and containers have been designed to stack one on top of another, but you must ensure that, wherever you keep them, they have a good air circulation between all the levels. Apples should be wrapped individually in newspaper before being packed into boxes – in the past it was held that pears should be packed into boxes in the same way, but without the addition of a newspaper covering, but now the view is that pears are wrapped similarly to apples; try some of each and see which method is the more appropriate in your own individual situation.

As for soft fruits, these are best stored frozen: raspberries, gooseberries, blackberries and black- and redcurrants all keep well in a freezer (as indeed, do blanched and sliced apples). You could also try drying apples and pears. Simply wash and slice the fruit and arrange the pieces in a single layer on a baking tray. Given that in Britain it is unlikely that weather conditions will permit the laying out of fruit in the sunshine over a period of days, it is probably best that one sets the oven to its lowest setting (130°C/250°F) and leave the tray for several hours until its contents have dried out and are almost crispy. Once dry, store in an airtight container and use within a few weeks.

Soft fruits will have to be prepared and stored frozen. Freeze in a single layer on a tray before bagging them up.

CHAPTER 4

Practical Poultry

Of all the livestock to be considered by the smallholder, poultry are one of the most suitable. They are relatively easy to keep, inexpensive to feed and house and have the added bonus that a willing friend or neighbour can look after them should your absence from home ever be necessary. It is unlikely that you will make your fortune from a poultry enterprise, but it is all a part of the whole, so to speak. Customers coming to your door for eggs may later wish to purchase fresh vegetables, and vice versa. It is also an important part of not literally putting all your eggs in one basket, but of spreading your risk, and your marketing strategy over as large an area as you can. Chickens and bantams are perhaps the most popular. They are very endearing and, although sometimes accused of being brainless, are really quite intelligent – they just appear to do silly things. If space is not a problem, chickens will prove the most profitable, but bantams really do come into their own when space is at a premium. However, the two types will happily co-exist. If rearing chicks is your aim, bantams often make

Silkies make good broodies, but silkie hybrids are, in the opinion of most people, even better. (Courtesy: Rupert Stephenson)

There are plenty of poultry books on the market, all of which will help to enhance one's knowledge. (Courtesy: Rupert Stephenson)

better mothers and will easily raise offspring from any type of poultry. It is not unknown, for example, for a Silkie bantam to successfully rear a couple of goslings, completely unperturbed at both the size of the eggs and the resultant chicks.

WHAT TO CHOOSE?

In poultry-keeping it is easy to get carried away with enthusiasm and end up with a smallholding full of back-to-back chicken runs, each containing a different breed chosen as a result of the heart ruling the head. Beware of assembling a collection of misfits and other's cast-offs – you are not running a sanctuary for retired poultry. Remember that it costs the same to feed an unproductive bird as it does a good laying hen and that you are in this enterprise to make a living, or to subsidize one. Be ruthless, once a bird has literally stopped earning its corn, despatch it to the pot and replace it with a productive one. This book is not about sentiment, it is to help you to survive as productively and economically as possible.

Whatever type of poultry is decided upon, it may be better for the novice to build up slowly with just a single breed and learn as much as he can about that one before diversifying into something completely different. One continually reads in the poultry and associated press where newcomers to poultry-keeping start out with a pen of one particular breed and have, in a short space of time, 'dabbled' with several others. If you are intending to breed under broody hens rather than in incubators, then, despite what has been written above, it does sometimes make sense to have a little mixed pen of nondescript breeds kept purely to act as hatchers and mothers of eggs and chicks from a breed in which you are particularly interested. Otherwise, however, it is better to be absolutely certain of the type of bird that really takes your fancy – buy a breeding pen of really good quality from a reputable breeder and learn all there is to know about its make-up, foibles and advantages. There can be no substitute for picking the brains of a local breeder or of a qualified judge of a certain poultry breed in which you are particularly interested; but, failing that, it pays to buy a copy of one of the many poultry-keeping books currently on the market. Tread cautiously and read wisely before being tempted too far into embarking on a hobby that has an undeniably wonderful way of taking over your life.

Chickens and Bantams

You cannot really consider yourself anywhere near self-sufficient unless you can collect eggs from your own poultry. In a small area and with only one or two human mouths to feed, two or three bantam hens might be the best option. Bantams take up far less space than large poultry, eat about a third of the amount of food and make less mess, and they also have a lot more character. The majority of smallholders keep poultry for their eggs and/or meat and so it is important to choose the right breeds for your particular enterprise. Some of the lighter Mediterranean ones such as Anconas and Leghorns will lay well and be less inclined to go broody, but, having said that, it is surprising just how many eggs some of the heavier bantam such as the Rhode Island Red, Light Sussex and Wyandottes will lay during the course of a year. Surplus males do not always produce a good table bird as they may have dark coloured skin, or, in some cases, do not fatten at all, producing a rather scrawny-looking carcass.

A Rhode Island Red cockerel crossed with Light Sussex hens will produce 'gold' or buff-coloured females and 'silver' or yellow males; the cross is often used by those interested in the practice of sex-linking. (Courtesy: Rupert Stephenson)

If table birds are your chosen option, then seek out a breeder who specializes in them. If you have an interest in breeding, either choose one of the traditional breeds and aim to sell hatching eggs or young birds to this specialist market, or, if you are fascinated by the possibility of sex-linking your birds so that the sexes can be readily identified on hatching by their colour, create a straightforward cross from a well known breed. For example, a Rhode Island Red cockerel running with a harem of Light Sussex hens will produce a wonderful, dual-purpose bird that is suitable for laying or meat. The resultant hybrids also have the advantage of being extremely hardy.

Light Sussex are a good, dual-purpose breed and can also be used when practising sex-linking matings.

LEFT: *There is a ready market for the hackle feathers of some breeds such as Old English game. (Courtesy: Rupert Stephenson)*

If you happen to have chosen some of the old English game birds or other rare breeds you may discover a useful and unexpected market. An enterprising man of our acquaintance makes money from the sale of the beautiful wing and tail feathers for the millinery trade and there is also a ready market for the cape feathers of the cockerels from trout and salmon fishermen, who are keen to buy these hackle feathers for fly-tying.

Ducks and Geese

Duck-keeping falls into two categories: the birds bred for the table or those intended for egg production. Although the demand for duck eggs is small in comparison with that for hens' eggs, there is a niche market. Some of the modern breeds such as the Khaki

ABOVE: *Free-range ducks add interest and character to any smallholding.*

RIGHT: *Geese are excellent producers of meat, grass mowers and guard 'dogs'.*

Campbell and the Indian Runner lay an amazing number of eggs in a season, while table ducks such as the Aylesbury may be mature and ready for killing at as early as eight to ten weeks. Ducks are generally messy creatures and, because of this fact, are definitely better kept free-range in order that they do not cause too much damage in one particular area. But having said that, free-range ducks on a smallholding will almost always be found dabbling around the base of any water trough set out for other livestock or quacking happily in and around a puddle caused by a leaking hosepipe or the latest deluge of rain.

Geese will lay eggs seasonally and their eggs make wonderful cakes and tarts, but the authors remain fully convinced that all goose eggs should be turned into more geese. They are wonderful birds and will get much of their living from a rough grass area or in the orchard under the fruit trees. With the addition of a small feed of hard grain twice each day they will thrive. Geese will also consume vast quantities of insect pests and have the added bonus that you will never need a guard dog. In Scotland many of the whisky distilleries are surrounded by grassed areas inhabited by flocks of geese through which no intruder has been known to approach unannounced. If you hatch and rear goslings in the early spring and these are then given a good supply of fattening ration a month or so before Christmas, they will become the table bird supreme. There has been, over the past few years, an ever-increasingly good market for them as an alternative to the ubiquitous turkey.

Turkeys

Traditionally, geese, fat cockerels and, latterly, turkeys, have always been bred for the Christmas market when many months of care and investment are realized in the sale of plucked and dressed birds. The money was important at a time of year when egg supplies were low and there was little other fresh produce to sell. The resultant income would then allow the farmer or smallholder to buy new stock in the spring and thus continue the cycle. Nowadays, there is a certain demand for turkeys on an all-year-round basis, but these are generally much smaller in size and weight than the Christmas bird which, in actual fact, has no upper weight limit because many of the heaviest birds are used within the catering trade. Until fairly recently, the popular trend has been for a white, broad-breasted turkey; but it is more likely that today's households will prefer a fresh, free-range Black Norfolk turkey as it fits in better with their mental picture of an old-fashioned, wholesome bird. These same people are prepared to pay a premium for what they want and will, it seems, travel to the farm or farmers' market in order to obtain one. It would be a foolish smallholder now considering turkey rearing who did not take full advantage of this fact.

PURCHASING POULTRY

There are many ways of acquiring poultry, but there is, unfortunately, only space here to discuss the main ones, together with a brief resume of their advantages and disadvantages. First, one can obtain birds from a friend or neighbour who keeps chickens. If the neighbour has raised his or her own check first that the parents were from unrelated bloodlines since to continually breed from the same line (that is, from brothers and sisters) can lead to some in-bred faults and this should be avoided. If the breeder has periodically obtained a cockerel from another bloodline then all should be well.

Markets, Sales and Auctions

Poultry can be bought at your local agricultural market, but the downside here is that the birds that are sent to market are often disposed of this way because they have become unprofitable through age or are the least profitable in terms of egg laying or meat production. You may get some good birds but it is very much a gamble. There is also a risk that you may possibly be buying in disease, in which case such purchases will be an expensive gamble. On the other hand, many poultry clubs and societies also hold sales and auctions at which fine examples of healthy and productive birds are sold off and, if the organizing group and the individual vendors are known to be reputable (as most of them are), one can generally buy with a degree of confidence. Many auctions and sales are advertised in the farming and smallholding press, as well as in your local newspaper.

Stock at various stages of growth from day-old chicks right up to point of lay can be bought from one of many reputable dealers and the choice will depend on how much you wish to pay. If the potential source is local to you and you are able to visit the breeder such a visit will serve to confirm the conditions in which the birds were raised; you will be able to inspect the parent stock and see the young birds and assess their quality before parting with your capital. Many breeders will then permit you to call and collect the birds yourself, thus ensuring their safe and humane transportation.

Buying Ex-layers

The final choice applies only when sourcing laying birds, but it is perhaps the most rewarding and cheapest option available to the smallholder. 'Battery' or 'intensive' hens are so called because they are housed in wire cages or kept in large numbers in intensive conditions. Young birds at the point of lay are housed just as they begin to lay their first egg and are kept until they cease to lay, perhaps a year from that date. At this time the bird will cease to lay for a few weeks, will moult its feathers and go into a resting phase before resuming laying. The farmer is not a heartless beast, but has only one aim which is to produce eggs, and so, to have to feed and house these birds during this period does not make economic sense. The whole batch of birds will be disposed of and replaced by another one on the point of laying and the egg production will continue almost uninterrupted. The discarded birds are mostly bound for the chicken pie and soup trade. If you are able to befriend such a farmer he will usually agree to sell you a quantity at a very good price.

The newly acquired hens will be a rather sad looking bunch, often with many missing feathers, particularly as they are about to moult. They will be unused to walking and will have been kept at a constant temperature. House them straight away in a well littered house with plenty of access to clean water and a little food. Reduce the light in order to prevent panic among them and ensure that the house is warm but with adequate ventilation. For a few days the birds will do little and eat little, but gradually they will become more active, they will begin to walk about and explore their surroundings and eventually will start to scratch in the deep litter, something they have never been able to do before. Gradually increase the light and ventilation until all seems well. On a mild day you can allow them to go outside and they will then quickly begin to behave as normal hens, scratching, dust bathing and

Keep in with the Neighbours

Before rushing out to buy your birds, it will pay to mention your intentions to the neighbours. It is essential that they are on your side as you do not want them to make any complaints to the local authorities or blame you for the appearance of rats in the garden. On that point, keep everywhere neat and tidy and do not leave food lying about. Keep sachets of rat poison under cover and in places where they might possibly try to set up home. If you have neighbours, consider carefully the idea of keeping a cockerel. At certain times of the year you will have more eggs than you need, even from two or three bantams. In the interests of good neighbourliness, make sure that you take them the odd half dozen – the gesture could have good results, with your neighbours offering you surplus produce from their gardens.

A movable house and run is perhaps the ideal way to keep a few head of poultry. Note the wheel system to aid easy mobility. (Courtesy: Rupert Stephenson)

chasing insects. Within a few weeks they should have regained their feathers and started to lay eggs. And these eggs are a premium, usually larger than in their first year and almost as many. With care, they should go on to lay for two or three seasons and need to be replaced only when they cease to lay altogether.

Resist the temptation to introduce a cockerel and breed from this type of bird. Their lineage is determined by careful cross-breeding to produce a specific type of hen destined to produce large numbers of eggs. By attempting to breed from them the results will be, to say the least, disappointing.

HOUSING

The one thing that saddens our hearts, and will, perhaps more importantly, sadden the hearts of potential customers, is a ramshackle poultry house leaning in the corner of a garden. Attached to the house is an inadequate run which is a mass of mud or bare soil, cluttered with old cabbage stalks and growing an impressive bunch of nettles in each corner. It takes but a little effort to make the whole enterprise look more humane and professional, your poultry will thrive and your results will reflect this.

With the threat of avian influenza, current suggestions for housing your flock are perhaps at variance with previous writings on the subject. The recommendation is that poultry should have sufficient space but with the capacity for them to be confined under cover should the authorities demand this in the event of an outbreak of bird flu in a particular region. If this is for a considerable period, it may cause problems with traditional free-range housing methods, which are designed only to house the birds at night. If they are confined in this housing for many days egg production will almost certainly cease, the birds may suffer from respiratory problems and even behavioural problems such as feather pecking or cannibalism may develop.

Arks and Runs

If you are intending only to keep and rear a few poultry, then you cannot do better than house them in small, movable units which can be periodically shifted on to fresh areas of grass. Sometimes known as the 'fold system', there is no better way of ensuring that your birds are kept secure from the unwanted attention of foxes and wandering dogs, that they have constant access to vegetation and grubs and insects and that they are unlikely to succumb to some of the diseases found in situations where birds are kept on one piece of land for a long period.

A number of arks and runs can be built or bought and they are equally useful in housing laying hens, fattening table birds, growers and a broody with her chicks. Even the smallest garden can accommodate a single unit containing a few bantams, and an orchard is the ideal place for a group of units. As an extra precaution and in an effort to prevent them from wandering too far once allowed out, they could be further protected by electric flexi-net that allows the inhabitants of a particular ark and run to be let out on one day and their neighbours the next.

The Balfour System

The housing we suggest for larger groups of birds is loosely based on the system pioneered by Lady Eve Balfour in the 1920s. Once

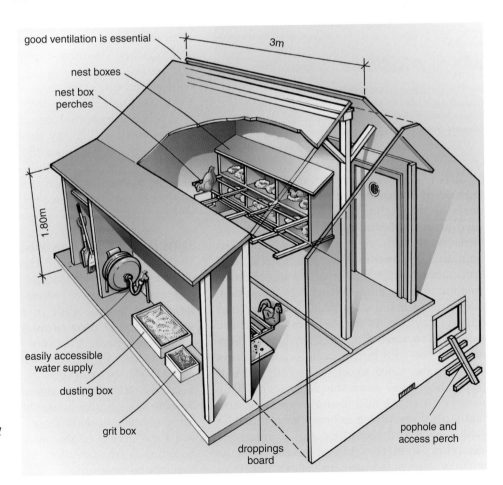

good ventilation is essential

3m

nest boxes

nest box perches

1.80m

easily accessible water supply

dusting box

grit box

droppings board

pophole and access perch

The ideal chicken house, which can be easily adapted to incorporate the Balfour system.

No matter what poultry is being kept, it is important that the pop-holes are large enough to allow birds an easy entrance and exit. (Courtesy: Rupert Stephenson)

established, it is very chicken-friendly, is organic in nature and easy to operate. The heart of the system is a well built poultry shed with sufficient room to house all the birds comfortably, rather than, as is so often seen, in a confined manner. The addition of a large covered run, either along one side or surrounding the house rather like a veranda, allows a plentiful supply of fresh air and natural daylight. It should be covered at the sides with a mesh size small enough to deny entry to the wild bird population; while this will involve more expense initially, it will prove beneficial over the long term. This covered area should contain feeders and water drinkers as well as boxes for grit and be roofed over with boarding and roofing felt extending from the chicken house itself or with a corrugated mineral sheet such as

Onduline. Not only will this avoid your feeding the local small bird population with expensive poultry food, it will, perhaps more importantly, lessen the risk of your stock contracting avian influenza, which is known to be spread via wild birds. In times when there is a perceived threat, your birds will be safely secured against risk and, with good management, you should notice little reduction in production. Their diet can be supplemented with greenstuffs brought in each day to the covered run (remember to clear the pecked and soiled greenstuffs each evening).

Depending on the land you have available, and dictated by its shape, a number of wire runs should be constructed, each served by a separate pop hole from this covered run. These small paddocks should be laid to grass, to which the birds are permitted access in rotation. Before the first run becomes sick with their droppings or turned into mud in wet weather, the birds are denied entry to it and permitted access to the next one. In due course, you can enter their old paddock, trim the grass and rake any droppings where they will be reincorporated into the soil. By rotating the stock in this way, the birds will always have fresh, clean grass and the paddocks can be rested for up to three or four weeks. During particularly wet and windy periods the birds can be housed in the main house and covered run.

Building a Chicken House
In the construction of the poultry house consider not only the needs of the occupants but also of yourself. Make it with enough headroom for an adult to stand up in and with an access door through which one can comfortably walk. Cleaning out will be made much easier for it. Consider how you are going to retrieve the eggs – you should be able to approach the nest boxes with ease from the outside of a smaller house; but, if the house is large enough to walk into, then ensure that you have access to the nest boxes and have both hands free to collect the eggs – juggling may not be your forte.

In the construction of your system consider the need for an 'isolation' ward. Without being hard-hearted, it is our experience that a sick chicken rarely recovers, but one that has a minor injury or has been bullied by the remainder of the flock, if given isolation and a chance to recover, may still serve as a useful member of the colony – a little TLC goes a long way. It makes good sense to isolate any sick bird until a diagnosis is made.

The general construction should be of good quality pine, both for the framing and the cladding, wood is naturally warm and, if treated, will last for many years. Make a point of coating each part with a proprietary waterproofing agent during construction as this will ensure that every joint and recess is well protected. After completion, give everything a further good coat or two, according to the manufacturer's instructions and leave it to dry out thoroughly before housing your stock. In an ideal world, the house should be given a good scrub down, inside and out once every year and a further coat of preservative applied. If this is not practicable, it should be carried out during any period when you are changing over your stock and before the new birds are introduced. If you have got your design right, and your construction is sound, the amount of labour required to keep your flock happy and productive will be minimal (for details on perches and nest boxes see fixtures and fittings in Chapter 7). While this system is primarily designed for chickens, it can be suitable for most types of poultry, particularly turkeys, with minimal adaptation.

All poultry houses, no matter what their size, should be well made and easily accessible to both the inhabitants and the smallholder.

Housing and Ponds for Ducks and Geese

Ducks and geese are not so fussy in their housing requirements, although it will still pay to enclose them in some form of run into which wild birds have little or no chance of entry. Because they do not perch, the housing for ducks and geese needs not to be as high as that for other types of poultry, but it does, however, need to be well-ventilated and easily accessible to the smallholder. Rather than a solid floor, it may pay to incorporate a strong wire mesh: small enough to prevent access by rats but large enough to allow faeces to drop through. Move the house periodically in order to prevent a build up of waste.

For ducks, a small pool can be constructed in the covered run area; they need access to an area of water for many of their activities, but it does not need to be especially large. Geese also relish a small pool and, if you take the trouble to surround it with an area of coarse pebbles or gravel, it will delay the area surrounding it from turning into one large mud bath.

FEEDING

The most labour-intensive method of feeding birds is the wet mash system in which a proprietary mash is mixed with water into a soft paste, along with boiled household scraps, such as vegetable peelings and stale bread. One of the authors remembers as a boy when he used to watch in fascination as his grandfather prepared this morning feast and hot, boiled potato peelings were 'pounded' in a wooden half-barrel with the use of two old table legs. Added to this was a measure of mash and a sprinkling of Carswood Poultry Spice. The hens and ducks devoured this with relish, particularly on cold winter mornings. The drawback to this system, picturesque though it is, is that it is highly labour-intensive, and additionally, any food not eaten immediately goes sour and is wasted. More

Free-range turkeys are traditionally housed in temporary straw shelters and appear to do well as a result. They must, of course, be protected, not only from the elements but also from predators.

importantly though, was the variability of the food value in the ration.

Modern mash or pelleted feed is far more simple and precise to feed. Self-feeders can be hung in the covered run area, high enough above the reach of rats and far enough from the sides to avoid spoilage from driving rain; the birds will help themselves to a balanced diet before roaming naturally outside and 'topping up' with fresh grass and the odd insect or two. It is amazing how much grass chickens will eat: if the new paddock is mown a day or two before they are allowed in, they will happily consume vast quantities, so never forget the importance of short, sweet grass.

Tasty vegetable waste and household scraps will obviously be offered to livestock as a welcome addition to their costly balancer diet. Tempting though it is to sling vegetables that have gone to seed or are otherwise unfit for the kitchen over the fence of the poultry run, it is far better to construct a rack to prevent wastage. An adapted garden hanging-basket or something similar is perfect for those with only a few chickens, as, hung at the right height, it offers exercise and interest to the birds as well as keeping everything out of the dirt. Move it around the run periodically so that no one particular place becomes overused.

As your poultry start to congregate before retiring for the night, it is good practice to throw them a few handfuls of hard grain such as wheat or cracked maize. This serves a dual purpose: it gives you a moment to inspect your flock and observe any signs of trouble; and, for the birds themselves, a cereal feed stays in the crop longer than the more easily digestible mash and pellets, a fact which is especially important, particularly during the cold winter months.

Feeders and Drinkers

On the subject of feeders, troughs and those wonderful water drinkers that can contain a couple of gallons (10ltr) of water which is automatically released into a circular trough

Feeding Kitchen Scraps

While there are very strict rules on the feeding of catering waste to any form of live-stock, it is possible to give your birds kitchen 'off-cuts' such as stale bread and the fresh leaves of vegetables, provided that they have not been in contact with meat. For example, it is perfectly in order to take bread or vegetables from the chopping board as you prepare your meal and feed them to poultry, but it would be illegal to take the same out of your kitchen waste bin if it also contained the remains of meat scraps. Also, beware of what you offer to the birds. One of us once accidentally killed a pen of his son's Black Wyandotte bantams by feeding them stale fruitcake, which then fermented in their crops.

at the bottom, a word of caution: there are some cheap, attractive-looking ones available in a variety of plastics, but it may pay to stretch the budget in order to consider a galvanized steel version. It will cost more at first but will stand a lifetime of use; it can be easily scoured clean when necessary and will withstand much daily wear and tear. Conversely, the plastic ones can be permanently damaged easily and cannot be repaired, they may be difficult to clean and may end by costing much more over the long term.

You could consider setting up an automatic watering system, which might go some way towards easing the burden for not only anyone looking after the birds, but also for yourself throughout the year. Apart from convenience, they are more hygienic than open-topped drinkers and will lessen the likelihood of disease occurring. There are several types to choose from, but whichever you decide on, make sure that you strip it down periodically, cleaning any valves and flushing through the hoses. It will also pay to keep a stock of spare valves, we have lost count of the number that have been dropped and lost in the grass. The 'trough' part, from which the

When considering the purchase of drinkers, it may well be economic to choose galvanized steel versions over the cheaper, plastic varieties. (Courtesy: Rupert Stephenson)

birds drink should, of course, be sponged or brushed regularly.

HATCHING AND REARING

If you are intending to enlarge your flock by raising your own birds from the existing stock then by all means go ahead, it can be very satisfying, and once a broody hen has sat on a clutch of fertile eggs she will do all the work for you. Keep her in a quiet, rat-proof environment and lift her off once a day to stretch and eat a little hard grain. Ensure that a supply of fresh water is near to her head and all will be well. Eventually she will walk away with perhaps a dozen fluffy chicks, which she will guard and supervise until they are big enough to be independent. She will soon let you know when she is ready to leave them. The two problems with this system are, first, that it is a relatively slow way to increase your colony and, secondly, you will end with only a few female birds to every hatching. The rest will be males that will have to be killed as soon as they can be sexed or kept and reared for meat. The next step up is to purchase day-old chicks, but again, if they are sold 'as hatched' they may contain equal numbers of male and female birds and this must be allowed for in your calculations. Buying sexed chicks is a better bet; allow for a (very) few mortalities and you should end

RIGHT: A broody hen is undoubtedly the best way of hatching and rearing the odd brood of chicks. (Courtesy: Rupert Stephenson)

BELOW: Rearing chicks from day-old obviously requires a heat source for the first three weeks. A hardboard surround will keep the chicks from wandering too far away from the heat in the initial stages.

with the required number of laying birds twenty to twenty-four weeks later.

To avoid the problems of raising day-old chicks (they obviously need some heat for the first few weeks of life and require careful handling), the next option is to purchase the birds at the age of five or six weeks. You would be unfortunate to suffer any losses at this age; they do not need heat and are truly independent. With good feeding and clean housing they should reward you with their first eggs in fifteen to twenty weeks' time. Alternatively, buy in your pullets at point-of-lay, they will be more expensive, but you will not have incurred the extra costs of rearing and feeding them expensive chick crumbs and growers' pellets. Timing is important and if you can purchase your new stock and time their start of the egg-laying cycle to coincide with the autumn, this will supply your enterprise with a fresh supply of eggs at a time when your existing birds are starting to lay fewer with the oncoming of winter.

A well-fenced run, plenty of grass and natural shelter offered by trees and hedges are all prerequisites of successful, practical poultry-keeping. (Courtesy: Rupert Stephenson)

MARKETING

With the growing awareness of the value of free-range, organic produce, your eggs should attract a premium. It may be worthwhile, if your flock is large enough, to invest in your own branded egg boxes as the customer will more readily identify with your brand and begin to ask for it. Selling eggs in this way may – depending on the volume of your weekly output – necessitate your conforming to certain legislative criteria whereby you will be required to date stamp and maybe even grade your eggs before being allowed to offer them for sale to the general public. Concentrate on quality: clean, even and well presented eggs are the aim. Goose and duck eggs are rather a niche market, some people do not like the stronger taste, others fear that the eggs may harbour disease. However, once having converted others to the delights of a duck egg, especially in cake-making, you may be able to build up a good market for them.

Little tricks of the trade will undoubtedly help to catch the eye of the customer. You may have noticed in the past that, when purchasing your free-range eggs from a local producer or at the farmers' market, there was always a small, clean feather or a wisp of soft hay in the box. Its inclusion may have been an accident or it may not. Noticing this on one occasion, it was casually mentioned to the seller who subsequently confided that it was his wife's enterprise and she had inadvertently put a feather in a box of eggs some years before. Apparently more than one customer had been delighted to see the feather; it made them think of the fresh farm origin and caused them to return for more on a regular basis. Since then his wife always took the trouble to do this, to show that they were not, in his words, 'from a factory'.

Raising birds from sittings on eggs or buying day-old chicks 'as hatched' will undoubt-edly result in unwanted cockerels. Or you may consider buying in a number of male chicks to fatten for meat. Whether by accident or design, sooner or later you will have to deal with them. The male birds should be separated as soon as their gender becomes obvious. The accommodation for these birds can be simpler than that required for laying stock, there will be no need for nest boxes, for example. A wind- and rain-proof shed connected to a covered outside run, coupled with a ready supply of good food and fresh water is all that is needed to produce a good table bird. Avoid the temptation to crowd them into a dark shed, you should be aiming for the free-range, organic market. Recent public awareness of the manner in which broiler birds are currently being kept has created a healthy market for birds raised in more humane conditions, and you, as a small-holder, will be well suited to supply this market in a small way.

Freshly collected eggs are always a desired commodity.

CHAPTER 5

Bee-Keeping

The first point to be considered is the very chapter title – you will be keeping bees, and the bees cannot be expected to keep you unless you enter into it on a very large scale. However, one or two hives will provide you with endless enjoyment. The bees' foraging will pollinate your fruit trees, and, with a little luck and a good season, they will reward you with a good personal supply of honey, long known to be the most pure and natural food you can obtain.

There are a few people who may suffer serious illness as a result of bee stings – a visit to your doctor will determine whether or not you are one of them. A few stings are always to be expected but will do you no long-lasting harm. Many bee-keepers go years without any stings, which are often the result of the careless or inconsiderate handling of these wonderful creatures. Bees interpret rapid movement near their hive, particularly when the hive is opened, as a threat, and will respond accordingly. When working with bees it is therefore important that one should always move in an unhurried and gentle fashion.

In the past bee-keeping was a rather murderous affair. A captured swarm of wild bees would be housed in a picturesque straw 'skep' until the autumn when the poor occupants would be destroyed in order to take their honey. With modern hives, the bees remain happy in their home for many years, and, by the careful manipulation the honey, if there is any spare, can be removed without harm to the bees. The bee-keeper may sometimes supplement their stores with sugar or 'candy' to see them safely through the winter months.

The word 'hive' has two meanings in bee-keeping. It is the name used for the structure in which the bee colony will live and is also the name used for the colony itself. It is important to remember that the honey bee (*Apis mellifera mellifera*) cannot survive as an individual. It is part of a colony, all the members of which serve the queen, who, in turn, depends on the colony for her survival. Experienced bee-keepers talk of the 'hive mind' or the way the hive behaves and the reasons for it. To learn about this and to observe it as a keeper of bees is very rewarding.

HIVES

The initial outlay can appear daunting. Hives, for example, are fairly expensive when bought new. However, if you join your local bee society you may find that its members occasionally have second-hand ones available (this is also true of most of the other equipment required). Properly maintained, hives will last for many years and there is little to fear from second-hand bargains, provided that they are sound and that they have been thoroughly cleaned and sterilized before use. By their very nature of construction, hive parts can be easily interchanged if they become unserviceable, even when occupied.

*ABOVE: When working with bees, one should always do so in a slow, unhurried manner.
(Courtesy: Steve Irish)*

BELOW: Straw 'skeps' were the forerunners of the modern beehive and were sometimes placed in a specially designed part of a walled garden so as to be near sources of pollen.

What Type?

The first decision to be made concerns the type of hive. There are several main sorts in use in the United Kingdom: the Smith, the National and the Langstroth being the most common. They have different capacities, but the Smith has been made more compact by making the frame ends shorter, the major advantage of this adaptation being that the hive is easier to transport at times when, for example, it is necessary to get the bees to the heather in season. For this reason they are popular in Scotland and the north of England. A further type is also available, the WBC hive (named after its inventor, William Broughton Carr) is the one often seen illustrated on honey jars. It has a similar capacity to other hives, but unlike them is double walled and is, in effect, a box within a box. Claimed by some to keep the bees warmer in winter and cooler in summer it is, unfortunately, a little more difficult to manage. Some, but not all, of the parts of the varying systems are interchangeable, but it is better to decide which single system to use and to stick to it.

How Hives Are Constructed

In essence, all hives work on the same principle. At the base is a floor, open at the front to permit the bees to enter and leave. On top of this sit one or more brood boxes, without a top or a bottom, each carrying a number of frames suspended on ledges. Each frame is fitted with a factory-made 'foundation comb' – a flat sheet of beeswax, sometimes reinforced with fine wire for strength. The bees will draw out this wax to form the familiar honeycombs, and these boxes are where the colony will live, raise their young and store some pollen and honey to feed the offspring. On top of the brood box go 'supers', which are more boxes, containing shallower frames. These are added as and when the brood frames are filled. They are divided from the brood chamber by a 'queen excluder', a mesh device that

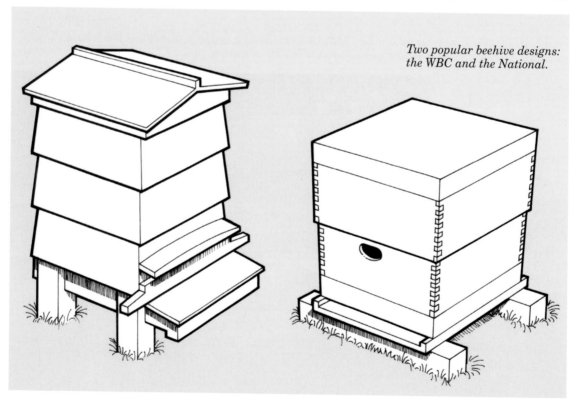

Two popular beehive designs: the WBC and the National.

Hives positioned directly under overhanging trees may suffer from undue dampness.

permits the workers access to deposit honey but prevents the slightly fatter queen from entering and laying eggs among the stored honey. In a good season, more supers may be added to a particularly active hive (depending on the honey flow) and they may be removed to harvest the honey as they are filled. The trick is in learning if and when to add supers and when to remove honey.

The top of the hive is covered by a crown board – a simple wooden panel with a shallow rim the depth of the bee space (the gap just wide enough for a bee to pass through). The board may have feeder holes, each roughly 2.5cm × 5cm (1in × 2in) and covered by a fine mesh panel to permit ventilation. When feed is required by the bees at any time, the feeder or candy block can be placed over the holes, the mesh having been first removed, allowing

the bees access to their food. Last of all comes the roof, often covered with zinc or aluminium to be totally watertight. There will be a small space between the sides of the roof and the walls of the hive to aid ventilation, vital to the health of the colony.

Siting Your Hives

It is important to prepare the proposed site with care and forethought. You will bring yourself unnecessary problems if, during bad weather, your hive stands start to sink into mud or the bees annoy neighbours with their activities. Once in place a hive should only be moved less than 3ft (1m) or more than 3 miles (5km), as the bees have an accurate navigating memory within their flying radius of 3 miles, and will return to an old site, whether the hive is there or not.

The site should be in the open, not under overhanging trees which will cause dampness. The preferred site should face east to get the benefit of the morning sun, but this may not always be possible. If there is a hedge or a wall to shelter the site from any prevailing wind, so much the better. A hedge is preferable as it permits the wind to permeate through more slowly and not roll over the top before coming back down in erratic gusts. The hives are best located at least 1m (39in) away for each metre of the height of the hedge to permit the bees space in which to fly up and over, to allow access for yourself and to avoid annoying the occupants if and when the hedge has to be trimmed.

Ensure that the chosen site is not near to a footpath or a neighbour's garden as bees fly in and out at frequent intervals during good weather and if their flight line is too close to people there may be accidental contact. Also ensure that the site is not too near your own or your neighbour's washing lines since bees

defecate as they leave the hive, even in winter months in fine weather, the sort of weather when washing may be hung out to dry. Newly laundered clothes covered with little yellow dots may cause some distress or annoyance.

The base of the hive should be firm and level. A concrete slab may appear to be ideal but may be ugly. Instead consider taking off the turf, making all level, putting down a layer of landscape fabric (bought from garden centres) and then laying a square of decorative paving slabs. This will be firm in all weathers, free draining and resistant to weeds from below. Bees do not enjoy a petrol-driven strimmer being operated close to their home. Finally, make the site large enough for a possible additional hive or two: if the bee-keeping bug really gets hold of you, you will be glad you did.

The hives should each stand on a firm base. Plastic beer crates are often used as they are strong and long-lasting, although they are not very aesthetically pleasing. Concrete blocks will do, but the best of all is a purpose-built, wooden stand.

ESSENTIAL EQUIPMENT

As well as choosing a site and a hive, you will need other items of equipment. First and foremost is the trade-mark bee suit: these are of two main types, the first is a jacket with either a separate hat or with a more solid veil attached; the second is the full suit, in effect, a white boiler suit with veil. The latter will give the best protection as there are no gaps where the bees can find access, and this factor will give the beginner greater confidence. Do you need gloves? Manipulating hive frames with special leather gloves gives safety from stings, but less sensitivity to the work. Many bee-keepers prefer to do without and chance the occasional sting.

All the right gear! Gloves, protective clothing and a hive tool are all essential tools of the trade for the bee-keeper. (Courtesy: Winchester & District Beekeepers)

You will definitely need a smoker. The copper or stainless steel models will last longest. A smoker consists of a container with a bellows and valve on one side. Many materials can provide the necessary smoking 'fuel'– newspaper; dry grass and wood shavings or hessian sacking are all successfully used by experienced bee-keepers. The object is to produce a cool supply of smoke over a period of half an hour or so when, for one reason or another, it may be necessary to have the hive open. It is well worth practising how to use a smoker before needing it for real; although not the end of the world, it can be disconcerting if your smoker has gone out when you need it. A few gentle puffs on the bellows now and then will keep it going.

The last bit of essential equipment is a hive tool. This is a simple piece of steel with a chisel-shaped blade at one end and a lever at the other. The lever is used to prise apart gently the frames which the bees have stuck together. The chisel end is used to remove the excess of wax or propolis (the resinous glue bees collect to fill gaps in the hive) from frames, lids and hive to permit them to fit back together snugly.

Useful, but not vital, is a pair of manipulating cloths, these serve to cover the top of the open hive while you are working on it, and, by progressively moving them, it allows you to lift out the frames in sequence and cover the areas you are not working on. This keeps the bees happy and stops them all from coming

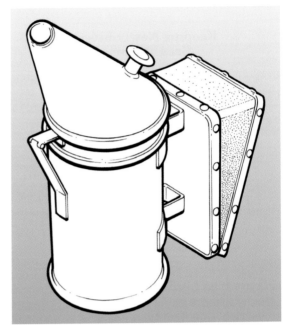

A smoker is an essential tool for all beekeepers.

up to the top to see what is happening. The cloths are available from your hive equipment supplier, but, having copied their design from a catalogue, you may be happy to make your own.

MAKING A START

You have decided on which type of hive you wish to use; each is clean, tidy and set up ready for the occupants. Now where do you get the bees? There are two suggested sources. Many hive manufacturers will sell you a 'nucleus', which is a starter hive containing a young mated queen and a number of attendant workers. They will arrive already established on frames to fit your hive (usually four or five of them) and will contain some brood, pollen and honey enough to last them until they establish themselves. The frames are removed from the box in which they arrive and placed in the middle of the brood box. The spaces on each

---

Coping with Stings

If you should happen to receive a sting when working the bees, simply take an unhurried step away, remove the sting, which will have been left in your skin (it will continue to pump venom for several minutes if it is left in). Then give the area of the sting a quick puff with your smoker as bees are attracted to the smell of a sting and the smoke will disguise the scent.

Keeping Newly-hived Bees at Home

A good tip to remember is to place a piece of greenery, a small branch from a shrub in a bucket of soil or even a potted shrub if you have one, a metre away from the hive entrance. This will stop the newly hived bees from flying straight out of their new home and getting lost. They will come out slowly and begin to make short orientation fights around the hive entrance and, after a day or so, the shrub can gradually be moved further away by stages and then removed.

side are then filled up with your new frames already fitted with foundation comb.

Another source of bees could be a member of your local bee society who has raised such a nucleus or hived a swarm. If he can assure you that they are good, clean stock from a suitable strain, this is a very good way to begin. He should establish whether your hive is at least 3 miles or more from his hives as, if it is less, the bees may exit their new home and fly back to their previous address due to their recognizing the area in which they have previously flown. At a greater distance than this the bees do not recognize their new location and will return to their new home.

It is not recommended that you take a swarm yourself without knowledge and experience. It may seem like a good opportunity, you may be able to take it successfully and you may be lucky with the result, but you will have no idea where the bees came from and they may already be bringing disease with them. They may also be of uncertain nature and, as a beginner, the last thing you want is a hive of difficult or aggressive bees. Also, if they have already swarmed, they may be from a strain that does so easily and will cause you problems in the future. It is always well worth the time and trouble to ascertain the nature of your bees before obtaining them.

WHO DOES WHAT IN THE BEE HIVE?

So, from one source or another, you are now the proud owner of a new colony. Having, we hope, done everything right before the bees arrived, they should now be left alone in order to acclimatize to their new environment. Whatever else you do, avoid the temptation to open the hive unnecessarily. The best way to see what is happening is to watch and observe – but not too close or you may get in the way of their flight line and inadvertently annoy them. Set up a seat a little way away and choose a warm, sunny morning for your observations. The bees should be out and about and, if all is well, there will be a number of bees flying in an apparently haphazard fashion in front of the hive. These will be young worker bees who, having spent their first few days of life on house-cleaning duties, are now orientating themselves as to the position of the hive in relation to nearby landmarks and the position of the sun. They will make progressively longer flights away from the hive, but will quickly return until such time as they are sure of their exact location. When the hive demands they will then begin to forage abroad, up to 3 miles and sometimes further, to supply the hive with nectar, pollen and water.

Once established, the flying workers should be arriving at frequent intervals from afar, and, if they are bearing pollen in the 'baskets' on their hind legs, it is a very good sign. Pollen is the protein or food of the hive. Nectar and, in time, honey, are the fuels they require. If they are bringing in pollen then the queen will be laying and the housekeepers will be feeding pollen to the brood and storing some for a rainy day.

From your observation point you should see the foraging bees returning and, from their frequency, you can gauge the strength of the honey flow. Spotting them returning with the pollen baskets on their back legs filled is relatively easy. Pollen can vary in colour from almost white, through yellows and reds to a

A fascinating sequence of photographs showing the retaking and rehoming of a known and trusted tame swarm. (Courtesy: Neil Vigers)

dark, brownish red depending on the flower source. Spotting them returning with a full 'tank' of nectar may take a little practice, but generally they tend to land heavily on the alighting board, frequently falling short and clambering laboriously up to the entrance. Often the bees will carry both pollen and honey, but if the hive 'mind' decides that they have sufficient pollen, workers returning with pollen only will be redirected and sent for the more valuable nectar. Flying workers also bring in propolis, which the house-

keepers will use to fill any small gaps in the hive walls, making it watertight and sealing everything, rather like a coat of varnish.

Inside the hive the young nurse bees will be busily occupied, cleaning out cells after the young have emerged, making new cells, feeding the larvae and, we hope, storing the incoming nectar and pollen. As the foraging bees return with their stomachs filled with nectar, the nurse bees will take it from them mouth-to-mouth. The nectar is deposited in the cells by the nurse bees. By the tempera-

Bees will travel as far afield as three miles in their search for pollen, especially when a perfect source such as this field planted for flower seeds is to be found.

ture inside the hive, the bees' own ventilation of the nectar and the draught of air created by the fanning of thousands of tiny wings, the water content of the nectar is reduced until the bees deem it to be 'ripe' enough to store. Each cell is then sealed over with a thin layer of wax.

When the nurse bees have spent about three weeks on hive duties they will begin to fly abroad and become foraging bees. They will continue at this task, flying many miles each day for about three weeks before dying. The slowly moving honey bee you sometimes see in your garden, with torn and tattered wings, is likely to be such a one. Worn out and unable to make any more flights, she will have given her short life for the hive. Spare a thought for her when you next have honey for tea.

As the day warms, a number of bees will be seen inside the entrance fanning their wings; this creates a current of air throughout the hive, maintaining a fresh environment and an even temperature. In an active hive, when the nectar is coming in at a great rate and the temperature is high, you may be able to smell honey wafting on the breeze, as many such workers draw out the warm air with the scent of honey on it.

You should also observe bees just inside the hive entrance. These are the guards – worker bees that check each new arrival by scent. It

Bees at work!
(Courtesy:
Steve Irish)

is their job to ensure that bees from other colonies, and, most importantly, their enemies, wasps, do not enter to rob the hive. It has been frequently observed that during times of plenty, when workers are bringing in large quantities of nectar, guards will permit entry to any bee provided that it is bearing nectar or pollen, but woe betide any foreign bee not fully laden.

When the hive begins to increase in size, the colony produces drones or male bees. They appear to live a magical existence. They do no work, they fetch and carry nothing, they have no sting so cannot protect the hive. They just eat and fly around on fine days awaiting the appearance of a virgin queen with whom they will attempt to mate. But also spare a thought for this poor soul – if he is successful in mating he dies, his genitalia torn out and embedded in his conquest. Those less successful will, come the autumn when they are no longer required, be unceremoniously evicted from the hive to die.

THE CAUSES OF SWARMING

The main task of the bee-keeper is to keep the colony happy and productive and to try to prevent the hive from swarming. As the colony behaves as a unit rather than a mass of individuals, the only way the hive can multiply is by part of the colony dividing and leaving for pastures new in the company of the old queen. The remainder of the old colony has a newly raised queen.

Swarm control, or, more accurately, swarm 'frustration', is the goal of the bee-keeper. If the hive swarms and the swarm is not caught in time, you will have lost a large proportion of your workforce and the chance of any honey surplus for that season. The reasons for swarming are many and varied, but only the main ones can be touched upon here. One cause is that the queen has become old or infirm. The colony is bound to the queen by a scent she distributes among the hive. When she is young and vigorous, this scent, passed right through the hive from worker to worker, keeps them all labouring away happily. They will make new combs to accommodate her prodigious egg-laying and she will respond to the incoming flow of nectar by laying more eggs as the flow increases. The colony will raise and feed her offspring, clean and prepare used brood comb for more eggs and store and prepare surplus nectar in the form of honey. As the queen's chemical signal becomes weaker the hive will quickly sense this and begin to make special cells, which are larger than normal worker bees' cells and often located in odd corners of the comb. The queen is induced to lay eggs in these cells (known, somewhat unimaginatively, as 'queen cells') and the workers supply these eggs with a different food, called 'royal jelly', which will turn the resulting offspring into a potential new queen. These cells may number up to a dozen or more in a hive. As soon as one of them is sealed over, the queen will decamp,

accompanied by a large number of her faithful followers, but not before they gorge themselves on the honey stores in order to prepare for the journey in search of a new home. Eventually, a new queen will emerge from the sealed cell. She may destroy her rival's cells then fly off to be mated before returning to continue business as usual. If the problem appears to be one of an ageing queen, observed by the declining rate of egg production, the cure may be to replace her with a new one. Mated queens can be purchased from reputable sources and, with some experience, the old queen removed and the new one introduced.

The other main reason for swarming is a lack of space – a vigorous colony needs plenty

Choose a warm, sunny day for inspecting the hives. (Courtesy: Winchester & District Beekeepers' Association / Russell Fairchild)

of room to expand, or at least the perception of it. If the colony outgrows its home, it can only divide off and seek alternative accommodation. The procedure and the result are the same: to avoid swarming, many bee-keepers follow a regular regime of frequent inspections, giving the bees room to expand and looking for queen cells, which they destroy as soon as they are noticed, thereby ensuring that the colony remains intact.

INSPECTION TECHNIQUES

We have mentioned earlier that bees do not like disturbance in and around the hives, but it is, however, necessary to make periodic inspections. Every week or ten days or so, during the season, it is essential to open up the hive and quietly and carefully go though each frame and generally observe the condition of the hive. Choose a warm, dry day when the hive is working well – around midday is good, although up to the early evening is acceptable. Ensure that there is no threat of thunderstorms as bees sense the lowered barometric pressure that accompanies storms and at such times even the most obliging of bees can become aggressive.

To begin these inspections, introduce a few light puffs from the smoker into the hive entrance and around the space under the lid. Fire is a great fear of the colony; when it is sensed it immediately reacts and the bees rush to the honey stores and fill their stomachs ready for flight. With a full stomach they become calmer and quieter. After a few minutes the roof may be removed and set to one side. Next, using the hive tool to break the seal around it, lift the crown board before blowing a further puff or two of smoke gently under it. Be careful not to overdo the smoke as too much can cause alarm and the opposite effect of that you wish to achieve – just a smell is enough.

After removing the crown board, a manipulating cloth is laid over almost all the top of the hive, while another should be rolled up and laid on the opposite side. As one removes each frame in turn, the rolled cloth is unrolled to cover the frames already examined and returned, the other is rolled up to permit access to the succeeding frames. This helps to prevent the bees from welling up on to the top of the frames and to keep them calm.

Having completed the inspection, the crown board is replaced, after the surfaces have been carefully scraped clean of any propolis you have disturbed in order to ensure a good fit. Early in the spring you would add a queen excluder and 'super', which, of course, you have previously set aside in anticipation. The crown board will then go on top of the 'super', and finally the roof should be gently replaced.

By checking the number of new eggs, the health of the raised brood and the quantities of nectar, honey and pollen stored, the newcomer to bee-keeping will learn much. It may take a while for the tyro to discover the queen for she is very elusive and will instinctively hide away (she may even be hidden by the workers themselves). If she is seen and appears in good health, then do not disturb

The Bee Dance

Foraging bees returning from a new and profitable source of nectar will tell the hive of its location by means of the wonderful bee dance. On the surface of the combs inside the hive the bees will perform a series of circles and 'wiggles'. The line of the wiggle will indicate the direction of the source of the nectar, the circle will indicate its distance and the vigour of the wiggle will convey information on the strength or quantity of the nectar available. It is also thought that other foraging bees will be aided by the smell of the nectar. This magical dance may often be observed during your inspections. So intent are they on transmitting the information to the hive that they will frequently continue the dance even when you remove the comb from the hive.

her. If all is well, replace everything as it was and close up the hive. The aim is to cause as little disturbance as possible, but, even so, your visit, no matter how brief and efficient, will result in production being disrupted for a day while the colony repairs the damage you have caused.

TAKING THE HONEY HARVEST

As the season progresses towards autumn and the supply of nectar begins to decline you will want to take your harvest. In a good year, if all has gone well, you may be fortunate enough to have several filled supers of honey on your hive. Now is the time that the hitherto benevolent bee-keeper turns robber.

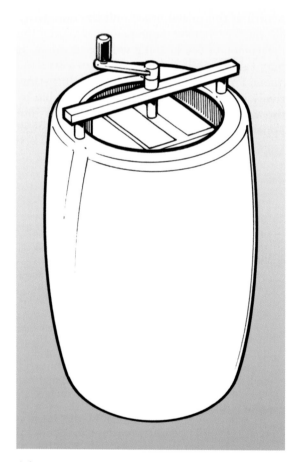

A honey extractor is essential, but can perhaps be borrowed from or shared with fellow beekeepers.

Warning!

Make sure that your chosen place is bee-proof! Tightly close all windows and doors and seal up any ventilators. Bees will follow the smell of honey and will try to take it back from you. A kitchen full of a few thousand bees will not make your task easy.

Remember, the colony members will not thank you for taking their bounty as it is their insurance against a time of need and their means of surviving through the winter. Good preparation is all-important. First you will require a honey extractor. This is a circular drum made of either stainless steel or food-grade plastic. Inside is a central axle around which are a number of wire frames that, when a handle on the top of the axle is turned, revolve around the axle. Many models have an electric motor fitted to make things easier. The frames of honey are placed around the axle, the lid is firmly fitted and, by the rapid rotation of the frames, the honey is spun out by centrifugal force, as is water in a washing machine during the spin cycle.

Frequently a honey extractor may be borrowed from a fellow bee-keeper or from your local society if you are a member. Fellow bee-keepers may assist you with your harvest in return for your help with theirs. This is a wonderful arrangement and, particularly if you are a beginner, can teach you much. The room allocated for this task, usually the kitchen, should be prepared in advance. Be sure to have on hand clean jars for the honey and a good knife with which to decapitate the combs, that is, to slice off thinly the protective wax covering all the ripe honey. A good supply of hot water for washing up is essential, as is an amazing quantity of jugs and funnels of all descriptions.

The precious honey should be allowed to stand in the warm for a day or so in order to permit any entrapped air to rise to the surface. Clean jars are then filled and, importantly, weighed and labelled. If you are intending to sell the honey, the label should clearly state the weight of the contents and the date of production, together with the name and address of the producer. Labels are important and may help to promote your product; they can be supplied ready printed by manufacturers of hives and equipment. Honey is a pure substance and if it is properly ripened before extraction and stored in clean, sealed jars it will last indefinitely. After a period in storage, and depending on the temperature, the honey may begin to granulate, that is, to form an opaque, solid form.

This is natural and the product is still fit for sale.

Having taken the honey, the (almost) emptied frames in the supers should be replaced in the hive. The bees will quickly clean out any remaining honey and take it down into the brood boxes out of your reach. After 24hr or so, the supers can be taken away and stored, sealed against insects, in a dry shed until next year. With good management, you will have taken the honey early enough so that there is still some nectar available to permit the deprived hive to replace much of its requirements for the winter.

WINTER FEEDING

It is about this time that you should assess the stores in the hive and consider feeding

Bees and foundation comb – the basic ingredients for a successful honey harvest. (Courtesy: Neil Vigers)

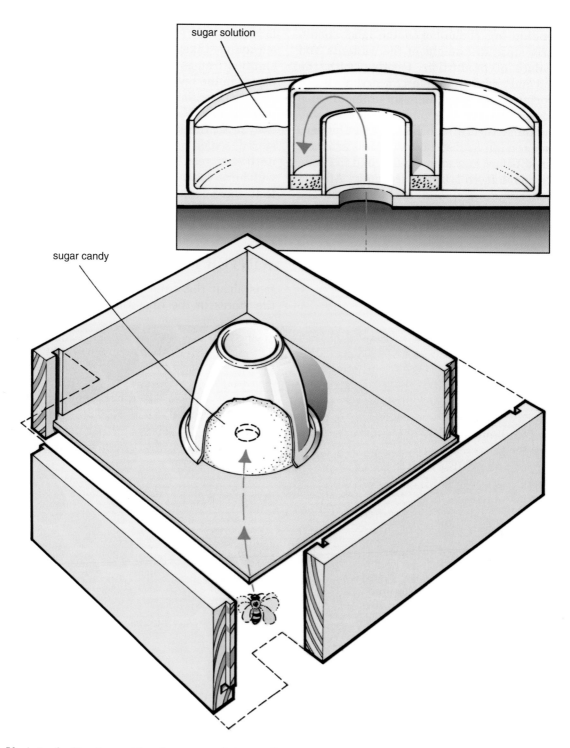

sugar solution

sugar candy

If winter feeding is considered necessary, it is usually done via an arrangement of this type, which is fitted on top of the crown board.

your bees to replace some of the stores you have removed. If there is not enough honey to last the winter the colony will starve. The object is to get the food to them as quickly as possible. If you begin to feed bees over a long period in relatively small amounts, the hive may misinterpret the supply. In response to the regular feeding, the queen could begin to lay more eggs to raise more bees to handle the supply. This would prove a disaster with the hive having more bees than it requires for its survival during the winter along with a consequential depletion of the stores.

The most commonly used feeder comprises a box of the same size as the hive and which fits neatly on top of the crown board. At the centre is a hole which is covered by a funnel-type device, which is itself covered by a plastic 'basin'. Once filled with sugar solution, the bees come up through the hole in order to take the sugar back down into the hive. Over 24 to 48hr they will take all they can store, after which the feeder should be removed and the hive closed up for the winter.

The sugar solution should be as concentrated as possible. Sugar is added to warm water until no more can be dissolved. If there is too weak a concentration, the bees will have problems getting rid of the extra water and the hive may become damp, which, in turn, can cause disease. During the winter months the hive will be a very different place from the summer hive. The bees themselves are different. The workers will live much longer, with a higher body-fat level to help to preserve them and are not destined for long-distance flight. They will spend the winter keeping themselves and the hive dry, clean and warm, and the queen will almost cease to lay, as they all await the return of spring. The bee-keeper's duties are merely to see that all is well. With experience, and by gently lifting the hive you can tell whether there is sufficient food in it by its weight. If you are unsure, or it appears light, all is not necessarily lost and a dry sugar substitute is available in the form of 'bee candy', a solid block of sugar contained in a plastic tub. You merely

remove its lid and invert it over the feeder holes in the crown board, after first taking off the ventilation mesh. The bees can help themselves if they require it and it does not contain any water. If it is unused it can be removed in the spring. Some bee-keepers refer to this as the bees' 'Christmas box', as this may be around the time that the bees need it. It is, in any case, good insurance against starvation.

All that now remains to do is to tuck the bees up for the winter. To prevent winter gales from lifting the roof, either place a heavy stone or a concrete block on top, or fit a strap under the hive and up both sides and over the roof. Proprietary ones are available or you may fabricate your own. Avoid the knotted rope approach – if you need to gain access to the hive during wet or frosty weather, the knot may prove difficult to undo.

On fine days during the winter, you may well observe bees flying. This will usually be to defecate or to fetch water and should not be confused with a resumption of activity. This will not take place until there is a definite change in the weather. The bees will soon respond to a rise in temperature and a lengthening day and all will begin again once more.

WINTER CHORES

During the winter months the bee-keeper should be preparing for the coming season. Used super or brood boxes should be cleaned and scraped free from any excessive build-up of propolis. They can be sterilized by gently passing the flame of a blow lamp over their surface. This will destroy any harmful creatures lurking in crevices. The outside of such boxes may be repainted if required, preferably with a safe preservative that allows the wood to breathe and is not harmful to the bees. The boxes should then be stored in the dry and covered to prevent insect attack.

Damaged comb in supers or a brood frame should be removed and replaced with fresh foundation sheets. The old comb may be

melted down and the wax recovered. Hive manufacturers will buy this from you to be reprocessed into foundation or you may wish to keep it to make such items as beeswax candles or even furniture polish to add to your income.

DISEASE

Once more it must be stressed that reading specialized books and up-to-date publications on the subject is essential. The fairly recent, accidental introduction of Varroa into the United Kingdom has caused many bee-keepers to reluctantly give up. Varroa is a tiny mite, just visible to the naked eye, it infests the hive and particularly the growing larvae and feeds on their body fluids. When the

weakened grubs emerge they are unable to support the colony. They will appear weak and stunted and, if untreated, the colony will die. It is estimated that over 60 per cent of the country's wild honey bee population has been killed by this pest. However, there are methods of control and it is ever more important that bee-keepers keep bees as, without 'domesticated' stocks to pollinate crops, there would be an ecological and economic disaster.

In brief, the main control used to be to introduce chemically-impregnated strips between the frames of the brood chamber after harvesting the honey. The medication, which kills varroa, but does no harm to the bees, is transmitted throughout the hive by the bees. Some weeks later these strips are removed, having done their work, and, it is hoped, leaving the hive pest-free for the coming spring. Sadly, there are already signs that the pest is becoming resistant to the treatment, but new methods are being developed. Bee-keepers now employ integrated pest management, which involves trapping and removing any drone brood that contains mites, and having open mesh doors to hives. This is a complex area that needs thorough learning and guidance.

In the USA, there were hopes that Colony Collapse Disorder (CCD) might have been a one-off problem in 2007. Unfortunately, this does not appear to be the case, with beekeepers reporting serious losses in spring 2008. CCD has the effect of bees either dying or disappearing completely, leaving only queens, eggs and a few immature workers in the hive. Various theories have been put forward as to the cause, ranging from the effects of insecticides intended for pests, mite infection and even the use of mobile phone handsets creating radiation and interfering with the bees' navigation system. The disease is thought in some quarters to be connected with Varroa although, at the time of writing, there is no evidence of this. However, researchers are experimenting with the possibility of 'foot-baths' impregnated with a newly discovered beneficial fungus being placed at the hive entrances. For the most up-to-date information contact your nearest DEFRA offices.

Two other bee diseases are European foul brood and American foul brood. The latter is the more serious. Once established, there is nothing else to be done other than to destroy the colony to prevent the disease from spreading. DEFRA employ a number of bee

inspectors in each area that you should call upon for advice. They are wonderful people, experienced and helpful. However, the numbers of inspectors are being reduced – a sad state of affairs just as they are needed most.

Despite the risks of bee diseases, the number of people keeping bees is, happily, increasing, perhaps due to greater public awareness of the environment. Once seen as the preserve of older men, bees are nowadays being kept by a wide and eclectic range of people. Thirty-somethings keep them on the balconies of city flats and business types keen to pursue a relaxing hobby are all joining the fast-growing band of apiarists. Suppliers of nuclei report they cannot keep up with the demand for new starter colonies. If you have the right temperament and a love of nature, do not hesitate to pursue this fascinating and rewarding occupation.

Regional bee inspector demonstrating to association members how to and what to look for when inspecting for diseases. (Courtesy: Winchester & District Beekeepers' Association / Russell Fairchild)

CHAPTER 6

Alternative Livestock Options

Depending on the size of the smallholding, it may be possible to consider more than just a pen of chickens or a hive of bees, but, if you consider alternative livestock options, it is important to choose possible breeds carefully. Modern farm animals are generally bred bigger in order to produce large volumes of meat, eggs and milk as quickly as possible and will, as a result, require more space. There are official requirements for space and equipment, details of which can be found in DEFRA's Codes of Recommendations, but it is well worth considering the options of rare breeds. Not only do they often require less space and are therefore more suited to the smallholder (as a rough guide, poultry need a quarter of an acre upwards; sheep and goats, half an acre or more and a Dexter cow, an acre), they have much more individual character than ordinary breeds of farm animals.

There is also the fact that one is helping to preserve a breed that might not otherwise survive due to modern-day farming practices. Different breeds of livestock vary genetically, with individual strengths and weaknesses, and it is probably the smallholding fraternity who have done more than anyone else to ensure the survival of some of the lesser-known breeds.

The Rare Breed Survival Trust can help and advise on all aspects of livestock care and management. It will assist you in finding a market for meat, the telephone number of your nearest sheep shearers and the location of any rams or bulls that might be required in order to service your animals. More importantly, they and DEFRA can help in ensuring that you do not fall foul of any legislation when it comes to movement licences, ear-tagging and cow 'passports'.

LEGISLATION

It is impossible to be specific in a chapter of this nature – which is merely intended to show what options may be suitable in your particular situation and the paperwork requirements necessary in order to comply with the legislation currently in force associated with the different types of livestock. It is, however, necessary to mention the fact that it is not usually just a simple matter of deciding that you would like a cow, a sheep, a goat or a pig on your smallholding and going out and getting one. Many of the rules may seem pointless, but they have been implemented for the well-being of the livestock and should, therefore, be understood and adhered to. Some are as a result of the recent outbreaks of foot and mouth disease (FMD) and much has been learned since the national disaster in 2001, lessons which undoubtedly helped in the more local outbreak in Surrey in the late summer–early autumn of 2007.

DEFRA often has a bad press, but, while admitting that we do not have to deal with them on a daily basis, we have found that on the rare occasions when advice and help have been required, it has always been possible been able to speak direct to someone who is

RIGHT: *A Dexter can be kept on an acre of land. (Courtesy: Stuart Webb)*

BELOW: *Some of the commercial breeds are out of the question for the smallholder.*

With more than fifty head of poultry it is advisable to check with DEFRA as to current legislation.

able to offer the definitive answer. For specific assistance regarding the legalities of bringing animals (poultry, unless you are intending to have more than fifty birds, are currently exempt from any form of registration) on to your land, you should contact either your local offices or check on the DEFRA website for the most up-to-date information. Remember that almost all regulations apply equally to a smallholder as they do a mega farmer.

Freedom of Movement

Assuming that you have registered your land with DEFRA and obtained a County Parish Holding (CPH) number (*see* Chapter 1, page 20) and, in the case of sheep, for example, acquired a flock mark (contact your local DEFRA animal health office for further details), you can, provided that the country is not currently in the middle of a livestock movement restriction (perhaps as a result of another outbreak of foot and mouth disease), consider bringing animals on to your land.

To do so, you will have to obtain a movement document (AML 1), available from your local trading standards office (contactable at your country council offices), or, more usually now, via the internet. Any livestock movements must also be recorded in a movement book, again available from either the trading standards or DEFRA. In it, it is necessary to record information such as the date, time of movement, where the animals have come from, where they are going to and the relevant CPH numbers of both places. In addition, the book will be expected to detail the fact that the necessary AML 1 has been completed and returned to the relevant authorities. And, on top of that, there are new animal transport regulations. It is not now possible to transport any animal further than 60km (37 miles) without first obtaining a special transport licence – for which you are required to sit an examination and attend a course. This makes transportation particularly difficult for those who do not have a local slaughterhouse or market close by.

Once livestock have been brought on to your premises, there is a six-day period whereby, with certain exceptions, a movement 'standstill' is enforced. The standstill period varies depending on whether you are moving pigs, sheep or cattle. Obviously any movement on to your holding requires the animal in question to be isolated from any others for the required period.

If you intend to keep cattle, it is essential that you register with the British Cattle Movement Service (BCMS), which is respon-sible for all cattle movements under the cattle-passport system. Contact their helpline on 0845 050 1234 and they will explain just what is required in order to register.

Ear-tagging

All animals require ear tagging, although the system for pigs and cattle varies slightly. The current tagging system for sheep, for example, requires every sheep to possess an individual, as well as a flock number, as a means of positive identification. Sheep must be tagged before they leave your premises or when they reach the age of nine months – whichever is the sooner. It is advisable to tag as soon as you can after any lambs are born; allowing enough space on the ear for growth, as, should an inspection take place and your sheep are found not to be tagged, the authorities are not going to be particularly happy.

New double-tagging rules were introduced as of January 2008, which, alongside a change in the requirements for movement licences and record-keeping, promise to ease

The stand-still period for pigs differs from that for sheep and cattle – not that it looks as if this particular Gloucester Old Spots is going anywhere in a hurry.

All sheep born after January 2008 now require 'double-tagging'.

the burden on farmers and smallholders by making the forms easier to complete. As far as the double-tagging changes are concerned (double-tagging has, in fact been required in Europe since July 2005, but the United Kingdom was fortunate in having a derogation, which, up until late 2007, allowed it to continue with the existing national system), it essentially means that sheep or goats born in the country on or after 11 January 2008 have to be double-tagged or have one tag and a tattoo, unless they are intended to be slaughtered before the age of twelve months, in which case they can continue to be single-tagged. Animals born before that date require only a single tag.

Bought-in sheep will obviously, therefore, already have ear tags that have been affixed by the previous owner. In turn, you are then expected to insert a second tag (as in the second place of residence). Any sheep that subsequently lose their tags are given a replacement tag. Tag records must be kept in a flock book, which shows the individual animal details, lambing information and any veterinary and/or drug administration facts and figures. Goats are treated in the same way as sheep.

Pigs are given a separate herd number for tagging or 'slap marking' (try slap marking a free-range pig …). It has to be done on both shoulders and must have been sufficiently well executed to be legible clearly at a slaughterhouse or market without any need to search through the bristles. Any pigs going to slaughter or market have to be tagged/slap marked whether they are under a year old or not. Pedigree herds still have notched ears as identification, and, if registered, the breed registration papers must be kept (for more on this, *see* page 105 in the section dealing specifically with pig-keeping).

Cattle require a passport with the correct herd number allocated to you and the dam's identification number. Two ear tags are required, one metal and one yellow (this colour changes if you lose the first and have to insert a replacement).

TB Testing

The State Veterinary Service (SVS) is a part of DEFRA and assists it in ensuring that animal health regulations are complied with. It is responsible for the compulsory testing of notifiable diseases, such as TB in cattle. These inspections vary in frequency, depending in what part of the country you live: some areas are inspected just once every four years, but, in locations where TB is known to be bad, an inspection will take place every year; you also have to have animals tested from those areas before any movement off farm. This was brought in after the 2001 FMD outbreak because, when restocking took place, TB was found in places where it had not previously been known. The SVS will give notification of when any testing is due and you will be responsible for making any necessary arrangements with your own vet. Failure to do so will result in movement restrictions (see above).

Medicine Records

There are requirements to keep a medicine book to record the dates and the types of medication administered. This is looked at by the trading standards officer who will probably visit once a year and, at the same time, will ask to see your movement book, copies of licences and, where applicable, even nowadays, horse 'passports'. You are also best advised to keep all receipts of animal feed purchased and records of where you obtained medicines, from agricultural suppliers, for example, if they have not been administered by a vet. It is also important to know the withdrawal periods of any substances administered to animals, especially before slaughter, this includes anti-fly treatment, dips, wormers and similar products.

CATTLE

Attractive though the idea of owning cattle may be, there are, as can be seen from the section on legislation above, one or two obstacles to overcome before you can do so.

However, if you fulfil all the criteria and have the housing, grazing and general space available on your smallholding – ideally, around 0.8ha (2 acres), then it might be a possibility to consider a house cow. Perhaps the best known breed of all such potential cows is the Jersey. They are generally quiet, docile and affectionate and will give more than enough milk to supply a household with cream, butter and cheese, as well as the daily pint. Obviously, she is not going to produce as much as the commercial breeds, but neither will she take up as much space nor demand as much in the way of hay and concentrates.

Smaller still is the Dexter: a charming, little, black- or red-coloured cow that can survive quite happily on about half the pasture required for a Jersey. It is extremely hardy and economical to keep and, unlike the Jersey is something of a dual-purpose animal, producing a fair amount of milk and providing excellent beef. Welsh Blacks are also small (but not as small as the Dexter). They too are hardy and make full use of poorer, hilly pasture, so, if your smallholding is of the hill farm type, it is a breed that may be well worth considering. Like the Welsh Black, the Shetland is a tough little cow, well suited to areas of poor grazing. It may appear dismissive to say so, but, although there are several other dual-purpose breeds, milking types and those bred specifically for meat, the majority of them are too large to be considered by the 'small' smallholder.

Whatever you choose, make sure that the animal you eventually buy is used to being handled and preferably led by a head collar and, if it is a milking beast, that it is also used to being hand- rather than machine-milked.

Getting Started

Depending on your ultimate aim, but assuming that it is milk from the cow and the possibility of slaughtering the calf for meat, you may decide to buy a calf (in which case, it will be roughly eighteen months before you can put her in calf), a heifer of between ten and eighteen months, an in-calf heifer, a down-

The Jersey (above) is probably the best-known breed of house-cow, but a Dexter (below) can survive quite happily on about half the acreage required for a Jersey. (Courtesy: The Jersey Society of the UK/ Adela Booth and Stuart Webb)

calver (a cow that has had a calf, is in full milk production, but the calf has been weaned away from her) or a mature cow (one that has already had two or three calves).

Remember that, because of modern-day regulations, it will be difficult to sell any surplus milk, and so, unless you decide to go into partnership with a like-minded neighbour and share not only the milk but also the responsibilities of ownership, you could leave the calf with its mother and purchase a second, unrelated calf that will also suckle on the cow. In that way, you can strip out some of the milk for yourself and leave some for the calves. When the time comes for her to dry off, the calves can, depending on their sex, either be fattened and sold for meat or sold as potential breeding stock. For this reason, there is much to be said for buying pedigree cattle at the outset, as the offspring will be more valuable and more easy to sell (to do this, your bloodline will have to be registered).

Breeding
A cow will indicate that she is ready to mate by 'bulling'. Typically, her milk yield will decrease and she will make an unusual amount of bellowing noises. Her vulva will swell and a discharge will be observed. As a smallholder, it is obviously impracticable to keep a bull and, as it is not going to be an easy matter to transport your cow to a suitable bull, the best way of putting her in calf is by means of artificial insemination (AI). It is possible to attend a course that teach you the methods of AI, but at around £450 to £500 for a three-day course, it is quite expensive and it may be better to call in 'the bull in the bowler hat', as AI inseminators were known when the practice was still in its infancy. Straws of cattle semen can be bought through local AI centres. Gestation is just over nine months (280–284 days).

Feeding
Good grass can, in theory at least, provide all the vitamins and nutrients required by cattle. Unfortunately, such a regime is impossible due to the fact that good grass does not grow all year round, so some form of supplementary feeding will definitely be required. In a

The offspring of pedigree stock will sell more easily. (Courtesy: The Jersey Society of the UK / Adela Booth)

A healthy, happy herd of in-calf Jerseys. (Courtesy: The Jersey Cattle Society)

particularly dry summer you may perhaps need to supply cattle with a little hay, but normally this would not be required until the winter months.

In order to help to maintain a house cow's weight and health, she must be fed 4.5–5.5kg (12–14lb) of best quality hay per day and 1.8kg (4lb) of properly balanced cake for every 4.5ltr (1gal) of milk she gives. This she should have in two equal parts during the morning and the evening milking time, cake first – hay later. Concentrates and hay will also be required for young stock. Your food supplier will recommend exactly what type is required. Vary the diet of cattle by the addi-

tion of cabbages, kale, turnips, carrots and other root crops, either grown in the vegetable plot or bought cheaply and stored carefully.

Health

Cattle can succumb to health problems and diseases similar to those that affect other types of livestock: scouring, bloat, worms and, of course, FMD. The potential problems of warble fly, ringworm, lice and mange can be lessened by the use of proprietary remedies available from your vet or agricultural supplier. As mentioned earlier in this chapter, it is essential that you should keep accurate notes and records of any drugs or preparations given.

A cow that has just calved may suffer with milk fever some two days afterwards. This is due to a sudden loss of blood calcium and must be rectified immediately by calcium injections administered by a veterinary surgeon. A cow may also come down with magnesium deficiency (known in some parts of the country as 'Herefordshire disease'), which, in its chronic form, may be fatal, but generally can be rectified by injections of magnesium, which must be administered very slowly. Mastitis is an inflammation of the udder caused by an invasion of micro-organisms by way of cuts, bruises, cold or dirt. Bacteria may also enter the udder through a teat injury. Typically, one of the quarters becomes hard and hot, and you will have a great deal of difficulty in trying to extract milk. It responds well to antibiotics, but remember that any milk obtained from the infected quarter must be disposed of safely, as it is infectious and that milk in the unaffected quarters will be affected by the treatment given.

GOATS

Probably the easiest way to start goat-keeping is to visit the goat section at a decent sized agricultural show (in fact, such shows are an invaluable means of examining any form of livestock), where you will see good specimens of most breeds and get an idea of what a good goat should look like. It also gives you the opportunity to make the acquaintance of experienced goat-breeders and there is almost bound to be a stall organized by one goat society or another, at which you can ask questions and seek advice.

As to the choice of breed, there are roughly four main ones kept in Britain today. The first and probably most easily recognizable is the Saanen – generally known as the British Saanen, despite its being originally imported from Switzerland via Holland. White in colour and placid in disposition, it is a good choice for the beginner. But so too are the Toggenburg and the British Alpine, the former tending to fawn in colour, with characteristic white marks on the face, legs and rump, while the Alpine has a black coat and white markings on the face and legs. For a really attractive goat you would be hard pushed to beat the Golden Guernsey, but that is only our opinion and there are devotees of the other breeds who would strongly disagree. Its colour ranges from deep cream to auburn, with mid-gold being considered the ideal by many. Apart from the Guernsey, all of the above originate from Swiss breeds, but one desert type well-known in here is the Anglo-Nubian, a strange looking beast with lop ears and a distinct Roman nose. Its colours are various, with roan and white predominating; it is heavier in appearance than the other breeds, and, although it undoubtedly milks well, some strains are, however, known to be extremely vocal which is possibly a disadvantage where any near neighbours are concerned.

Although the Saanen, the Toggenburg and the Alpine may have the prefix of 'British', there is also a goat simply known as the British – in reality, it is a hybrid and the result of crossing two pedigrees in order to create certain desirable genetic characteristics. Examples can vary greatly, in some they show the looks of one parent, while in others they look like neither one nor the other.

Getting Started

As a newcomer to goat keeping, it might be best to buy a female that is in kid and still milking from her previous kidding, as animals do better when they have kidded with you than when they are moved soon after coming into milk. A young female which has been mated is probably the best option of all, as she has the whole of her productive life of ten or more years ahead of her.

Should you decide to buy a goat that is already in milk, you should obviously learn to milk before acquiring her. Goats are easier to milk than cows, and milking is a knack which

Plentiful grazing leads to contented animals.

is quickly mastered, although it should be noted that it is sometimes difficult for the teat of a goat to accommodate all four fingers of a male hand. As in all situations such as this, it is best if you can get someone experienced to show you how it is done. Goats need to be milked twice daily at as near to 12 hourly intervals as possible. Do not expect milking to be without mishap the first few times – you will be fortunate if you do not end up with the goat lifting her leg in protest at your clumsiness and knocking over the bucket in the process.

Breeding

To ensure a continual supply of milk your nanny will need to be mated – which must be done in the 'rutting' season, from September until around mid-February. She will come into season every twenty-one days for most of these five months. If you see a constantly wagging tail, look at the vulva, which is slightly sticky, swollen and pink when a nanny is in season. Once her cycle is known, arrange either to take her to a billy, or, if you are unable to find one of the required breed, try and find out about the possibilities of artificial insemination, either through your local smallholding club or your veterinary surgeon. The nanny should be checked twenty-one and forty-two days after mating to see whether she is not in season. After this time, she can be assumed to be in kid. As the goat, like the sheep, carries its young for five months,

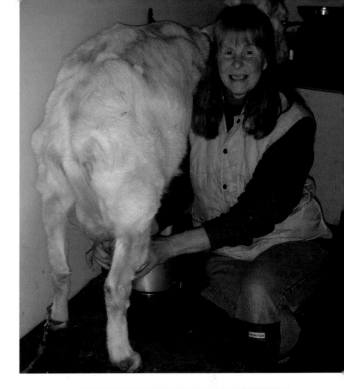

Milking is a skill that is soon perfected, but women are often better at it than men due to their greater patience and smaller hands!

October and November are ideal mating times, as the kids will be born in April and May at a time when, you hope, the weather will be improving and they can be expected to do well as youngsters.

Feeding

In summer, when they can forage for themselves, goats need little extra feed, but in winter, when they have to be kept in their shed for all or part of the day, some hay will be appreciated along with a regular scoopful of concentrates. It is advisable to buy a ready-made goat ration from an agricultural feed supplier because it will have been carefully and scientifically balanced in order to supply the right vitamins and nutrients; it takes the guesswork out of mixing your own (traditionally, a blend of crushed oats, flaked maize, bran and linseed). Goats will thrive on vegetables such as kale, swede and turnip and can be given the outer leaves of most greens before the crop is taken into the kitchen. They also like dried, stale bread.

Being somewhat indiscriminate eaters, goats tend to try grazing on things that other animals would have the sense to leave alone, including poisonous plants such as yew, rhododendron, laurel, laburnum and box. Ensure that such plants are well out of reach. Incidentally, garlic is well known for its medicinal and health properties. Highly antiseptic, it is rich in sulphur and is also one of the best natural worm expellers. It also helps to immunize against infectious diseases and in treating gastric disorders and rheumatism, as well as being effective against parasites such as ticks and lice. It is also said to increase the fertility of goats and, indeed, of most animals. Marvellous stuff, but beware of the fact that, given frequently and in too great a quantity, it is likely to taint the milk.

Tethering Goats

Goats are inquisitive and know that the grass (or bush or tree) is always greener on the other side of the fence. To prevent their climbing out of their enclosure, use stock fencing as described in Chapter 7. In some situations it may be necessary to confine a goat by means of a chain, a wide collar and a free-running metal ring to prevent wear and tear on the collar. The tethering peg (and, for preference, also the chain itself) should have a swivel joint to prevent the chain from becoming tangled or knotted. Tethering goats in this way is not a practice recommended by ourselves, but if it has to be done it is important to observe the following:

- Move the tethering peg at least once a day.
- Be sure that the grazing area chosen is safe from dogs that may attack a tethered goat.
- If you have more than one goat tied in this fashion; ensure that they cannot reach each other or they could choke each other with their chains.
- On hot, sunny days, make sure that the goats are able to find some shelter but that they cannot get right around trees or bushes.

Health

Much of what is written below regarding the health of sheep is also relevant to goats. As well as a proper programme of prevention regarding worms and external parasites, goats too need a regular foot trimming routine. It is also a good idea to give their hooves a regular treatment of linseed oil or proprietary horse-hoof oil as this will help to keep them in good condition, especially during a hot summer when the hooves may otherwise dry out and crack.

Watch out for bloat and scouring (both self-explanatory from their names), which can be caused by a goat's eating too much fresh greenstuffs or wet, frosted or lush grass. General signs that all may not be well are a loss of appetite, a dull, 'staring' coat, a change in the texture of faeces, failure to chew the cud, a cough, lameness or, in the case of a milker, a drop in her yield. Seek the advice of a vet if the condition persists for more than a day or two, a recommendation which, in fact, holds true no matter which livestock appear to be suffering.

SHEEP

As well as providing meat and wool (and, should you be so inclined, milk), sheep are perfect for grazing orchards and paddocks, as they cause hardly any damage to pasture, graze it cleanly and efficiently and manure it well. You may think that a sheep is a sheep, but there are several breeds to choose from, some more suitable for the smallholder than others. Although they are a rare breed, Dartmoors are considered by many to be a good choice; Jacobs are another, but, as the National Sheep Association lists over eighty breed societies, there is obviously too much of a selection for us to begin discussing specific

A young, healthy goat should possess a long neck and a clean throat. The body and ribcage area must be deep and the rump gently sloping.

RIGHT: Choose your breeds carefully as some are suited to rough pasture and moorland ...

BELOW: ... while others will only really thrive in more forgiving conditions.

breeds here. One ancient breed that is worth a mention, however, is the Castlemilk Moorit now classified as 'endangered', their meat is more like venison than lamb; they lamb easily and are full of character. Their only disadvantage (and it is a serious one) appears to be that, although they are resilient to illness, they are, unfortunately, susceptible to the disease known as scrapie. If recent governments had had their way, the breed may have eventually died out due to the fact that the National Scrapie Plan aimed to improve resistance to the disease by interbreeding with only the less vulnerable breeds. Interestingly, as soon as the Rare Breed Survival Trust began making would-be sheep owners aware of the National Scrapie Plan, the sales of Castlemilk Moorits rocketed!

Whatever type you eventually choose, it is best to start modestly; but you will, in all probability, be restricted as to numbers by the amount of grazing at your disposal. One hectare (2.2 acres) of land will support around eight sheep, plus their lambs. If you want to overwinter your animals without too much supplementary feeding, you will need either twice as much land or half the number of sheep. Sheep are quite heavy grazers and should be periodically moved from one paddock to another. The paddocks will benefit from being split into four: sheep, like other grazing animals, always appreciate a change of grass. In the winter, sheep should have access to shelter, and will need a supplementary daily feed of concentrates, plus hay and any surplus root crops from your vegetable patch.

Sheep must be sheared annually in the spring or the early summer. With only a few animals, it might be possible to shear by hand. Easier though, is to find a sheep farmer and/or breeder who will do it for you. If you cannot find such a paragon, seek out the services of a contract shearer who is prepared to clip small flocks. Shearers normally charge per sheep, but the actual price will probably vary according to the distance they have to travel.

Making a Start

It may be best to start with in-lamb ewes, which means buying in the autumn or early winter. It is then a matter of finding someone reputable who has animals available; try to purchase from someone with a small flock, as you are more likely then to obtain better stock than you would by going to a purely commercial breeder. The best ewe to buy is one around two or three years old that has already had a couple of lambs. Where possible, take someone experienced with you and watch how he or she assesses a good quality animal so that you have some idea as to what to look for the next time you go to buy. Generally, they will feel through the wool in search of a broad back, inspect the sheep's mouth and teeth, and look for a bright eye and a clean 'back end'.

Feeding

Sheep can survive on little more than good pasture for most of the year, but it is essential that they receive some good, high protein supplement in the late autumn and winter. A salt block will always be appreciated, as will some turnips, swedes, kale and cabbage, if there is ever any surplus in the vegetable patch. Rather than feed on the ground, sheep should be given access to a hayrack, feed trough and, of course, a plentiful supply of clean, fresh water.

Lambing and Breeding

Your in-lamb ewes will benefit from being given a dry shed or barn around the time they give birth (twenty-one weeks after mating). Each ewe will, depending on its age, give birth to one, two or three lambs, which should stay with the mother until at least shearing time.

For lambs in subsequent years, you will obviously need to find a ram. Whether you buy or borrow one depends on whether you have friendly sheep-keeping neighbours, although the problem in relying on a borrowed ram is that its owner will not want to lend it to you until he is sure that it has mated with all the ewes in his own flock – which will probably mean that your lambs will be born late. If you choose to buy a ram, remember that he will be sexually mature at anywhere between seven and fifteen months and that ewes are in season several times a year, so you should keep the ram and ewes separate until you want them to mate. Having decided at what time of year you wish to lamb (usually January, February or March), work backwards five months (the gestation period) and add a further fortnight, that is the time at which you will need to introduce your ram to the ewes. A ewe is capable of producing two or three lambs each year until it is around the age of fifteen, but her fertility will begin to drop after the age of around eight.

Fattening and Castration

It is usual on a smallholding to keep the female lambs or to sell them for breeding and to send the males to the abattoir. They will fatten quite well simply by having access to good pasture or by being on a diet of hay, concentrates and a few surplus vegetables, as indicated earlier. The length of time necessary for fattening varies. Young lambs make the tenderest meat and are traditionally killed around Easter, but it is more usual for the majority of lambs to be brought on until the autumn.

To obtain the highest quality meat, you should aim to have the male lambs castrated as early as possible. Castration can be done in two ways; one is by the use of an 'elastrator' and rubber rings (which should be done at one or two days' old, in order to cause the least amount of discomfort), the other is to cut the sperm cords by using castrating pliers – the latter method can be done at any time during the first four weeks of life, but after this it must be done only by a vet. Although castration is not absolutely necessary, it will help in producing a better carcass and allow you to run both young males and females together – an important factor when space is limited.

Health

Sheep do tend to suffer from a number of problems, one of the most common being worms, which are picked up when grazing. Worms can live inside a sheep without being noticed and most shepherds treat their sheep with a wormer as a matter of routine; but it is possible to minimize the risk of reinfection by moving sheep from one pasture to another every two weeks and not returning them to the first pasture until a month has passed.

External parasites such as ticks, maggots and ringworm can be controlled and it is recommended that to do so sheep should be treated with a suitable deterrent three to four times a year (it is no longer a legal requirement to dip).

Occasionally, sheep go lame due to inflammation of the joints. For mild infections use a gentian violet spray or a mild iodine solution. They are essential to the smallholder as they can be used to treat cuts and wounds on all types of animal.

About every six weeks, sheep, like goats and cattle, require a pedicure. This is best done by turning the sheep over into the same position as if for shearing. By using a sharp knife, pare off a small piece at a time and bring the outer horn flush with the sole of the foot. Hooves that have been neglected for a time may require to be cut back with special hoof shears, available at any agricultural suppliers.

PIGS

Pigs may have their place on the smallholding, but, without much space for them to root about and graze, it will probably be necessary to keep them in an old-fashioned sty or contained by an electric fence in a wooded area or orchard. If you keep pigs on a small patch of land but without good care and a regular rotation, they will ruin it and it will be the devil's own job to bring it back into good condition. All land is too valuable to be out of production for any length of time and this is especially true of that belonging to the smallholder.

If you value your land, it may be best to keep free-range pigs in a wooded area.

Breeds

The Gloucestershire Old Spots is a large, lop-eared breed and is a good all-rounder: it provides tender, succulent, fine-grained meat that is particularly suitable for pork or bacon production, it is an extremely hardy breed and, in most parts of the country, it can be kept out of doors all the year round, but provide an ark for farrowing. Finally, Old Spots sows make excellent mothers, as they are docile and easily managed, they also continue to rear good litters long after those of other breeds have reached the end of their breeding life. Large Blacks are one of Britain's oldest breeds and had their origins in the English hog of the late sixteenth–early seventeenth century. Like the Gloucester Old Spots, they are extremely docile, hardy and suited to simple outdoor systems; the sows are excellent mothers, with exceptional milking ability, and are able to rear sizable litters on unsophisticated rations.

Other breeds include Large Whites, Middle Whites, Landrace, Tamworth, Berkshire, British Lops, the Oxford Sandy and Black, and Saddlebacks; some of these make a better choice for the smallholder than do others. Contact the British Pig Association and take advice from them as to a particular breed's suitability and characteristics.

Feeding

With many of the old British breeds now enjoying something of a revival in popularity (due in no small part to the efforts of organizations such as the Rare Breeds Survival Trust and specific pig breeders' clubs), it pays

to take advantage of their availability, not least because there is often not the necessity to feed them expensive concentrates exclusively, as would be the case with commercially-produced animals.

When allowed to graze, the diet can be supplemented with potatoes, roots and windfalls. A natural diet may take longer to produce meat, as such feeding is not necessarily as efficient as specialist pig foods and the pigs will use up energy snuffling about and in keeping warm. The smallholding may provide food in the form of surplus vegetables or in milk from a goat, but it is perhaps best not to be tempted into giving them kitchen scraps in case you fall foul of legislation which stipulates that it is illegal to provide a diet of 'scraps' that may include meat waste; it has been proved that such a diet has, on occasion, been the source of swine fever and FMD.

Supplement the diet of free- or semi-free-range pigs with root vegetables, windfalls and a few concentrates.

Not two-tone pigs, but merely the result of a recent wallow! Pigs need to protect themselves from sunburn and mud is their preferred way.

Unless your pig is completely free-ranging (very unlikely on a smallholding), it should be fed some concentrates.

Breeding

As with sheep and goats, you will find it easiest to start your pig-breeding enterprise by buying a pregnant sow (females may begin to produce at as young as six months). Subsequently, you can mate her with a nearby boar or use artificial insemination. Members of the British Pig Association (BPA) or local pig breeders' clubs can help you with more details.

Signs that a pig is in season – which can be at any time of the year – include its standing perfectly still, a deep grunt and a reddening of the vulva. If mating by the 'natural' method, try to find a boar owner who is willing to keep your sow for a few days when she is in season, as, otherwise, you may make a journey to the boar only to find that the female is not ready and must wait another day or two.

The gestation period is nearly four months – three months, three weeks and three days, to be exact – and you should allow a sow two months for suckling her litter. After weaning, she can be in pig again within a week or two, so you could, in theory, have two litters (and up to thirty piglets) in a year.

Registering Pedigree Stock

Once you own a pure-bred pig and should you wish to register it for breeding purposes, you must join the British Pig Association. You may offer your own selection of herd prefix for their approval and you will be allocated herd destination letters, from which all pigs bred and registered by you may be recognized (this is in addition to the ones required by DEFRA for normal ear-tagging procedures).

Make sure that the pig you wish to register is either herd-book registered or is otherwise eligible. If it is already eight or more weeks old it must have a legible number tattooed in its ear. It must also be a member of a litter whose birth, ear numbers and parentage details have been submitted to the BPA, and this, plus the necessary ear marking, can only be done by the breeder (registration can take place at any time, but the breeder will be required to sign the application for registration form).

HORSES AND DONKEYS

It is very unlikely that, unless you have the knowledge, interest and at least a fundamental understanding of horses, you will be tempted into 'wasting' valuable productive space on your smallholding by considering the acquisition of such a beast. Furthermore, it is our view that, unless you intend to compete or hunt with horses, they are too time-consuming for the average smallholder. They require a great deal of money to be spent on them – on good quality concentrates, hay, grooming, shoeing, stabling (or, at the very least, some form of field shelter) and also the regular, daily regime of picking up droppings from the pasture. It is a sweeping generalization, but the quality of much grazing land in the United Kingdom has deteriorated as the number of horses being kept has increased. One only has to look at the average horse paddock to see ragwort (which they

Donkeys are less demanding than horses and ponies in terms of time and money, but ask yourself why you actually want one?

A veterinary passport is required for horses and ponies; irrespective of the animal's size.

will, thankfully, not touch all the time it is living) and other weeds growing around the soured areas chosen as 'staling' (urinating) and defecating places by the occupants to see the proof of this statement. Unless the droppings are picked up regularly, horses will not graze in the immediate location and the field soon becomes an untidy mess of straggly grass and patches bitten bare.

In all cases, a veterinary 'passport' is required. On it can be found the height, breed, sex and age of the animal as well as several pre-printed profile drawings that help to identify any distinguishing marks or features.

Donkeys

We having rather brutally (and probably unfairly) dismissed horses in a short paragraph, it is, despite their being a member of the *Equidae* or horse family, perhaps a more practical proposition to suggest that the smallholder consider the purchase of a donkey. It will do well on poor grazing, in an orchard maybe, provided that it cannot get anywhere near the trees (the fruits of which are not good for it in quantity, especially when they have fallen to the ground and begun to ferment), which, unless they are fenced, the animal will ring bark in a short space of time. Donkeys are also less demand-

Donkeys as Companions

Most animals thrive on companionship, even if it is not with one of their own kind, and one should consider carefully before deciding to purchase any single animal. Indeed some, such as donkeys, can be a calming influence in situations where other livestock is kept.

ing, both in terms of time and money, but, before you go ahead, ask yourself why you actually want one. Three or four years ago it was fashionable for newcomers to the countryside to have a donkey in much the same way that, a few years earlier, certain breeds of pig were the 'in thing' to have. Bought on a whim and without any realization of the care they require, it was not long before donkeys were being given away or sent to animal

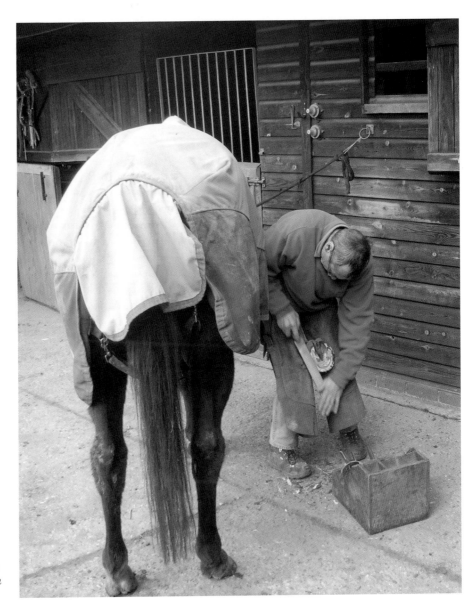

All equines require periodic visits from the farrier.

rescue centres. The fashion for donkeys has continued, so it is nowadays a relatively easy matter to get hold of one, should you feel the need to supply the grandchildren with a pet or as company for a single goat, for example.

A complete fallacy seems to have grown up around donkeys: that they are as hard as nails and grow fat in a desert; in fact, donkeys are nowhere near as hardy as the smaller of our mountain and moorland ponies, which, after all, are native to our climate. Although donkeys do not need the same care and attention as their horse cousins, they still require some form of shelter (they do not like or do well in wet conditions) and secure fencing. They also need regular worming and periodic attention from the farrier.

It matters not whether you intend riding or driving your donkey, even standing 'idle', its feet will still require attention from the farrier every three or four months in order to prevent them from growing out of shape or developing what are known as 'sandcracks'.

For ease of management, donkeys need to be docile and tame. Small though they may appear, they are still a lot stronger than you or us, not for nothing do we sometimes refer to someone as being as stubborn as a donkey. Where possible, buy a young animal that you can train, or, if choosing an older one, make sure that it is easy to lead on a head collar and has no objection to being groomed or having its feet picked up.

The sex is also important: a castrated male is going to be far easier to handle once fully grown than is a stallion. A mare has the obvious advantage of being able to be bred from, but comes into season whenever its hormones initiate the pattern of regular periods of oestrus (mainly in the spring and summer) and can develop traits such as nipping or being skittish, which are out of character and not normally seen at other times of the year. Breeds are mainly defined by size; living where we do in France, the emblem of our particular department is a Poitou Charente donkey, as big as a pony and distinguished by its (intentionally) matted coat. At one time, it

Blacksmith or Farrier?

There is often confusion between the terms blacksmith and farrier. At one time, the country blacksmith would have shoed horses, mended farm implements and constructed any sort of wrought iron work for whoever required it. Now the name blacksmith is given only to those who specialize in the wrought iron work element. A farrier is one who tends to only a horse's feet, making and shaping shoes and trimming off hooves where necessary. He will have had a long and hard apprenticeship before being allowed to work with horses and ponies, and his understanding of the legs and feet of an animal is on a par with that of a veterinary surgeon.

would have been seen around every rural dwelling in this area but it is now rare, there being only a couple of hundred left in the world. Others are more common, ranging from the Miniature Mediterranean, standing less than 1m (3ft) at the withers, to the type known as the Jackstock, which originated in the USA. In between are the breeds developed for work in hot countries other than the Mediterranean, which, although still well-known in their own environment, are more unusual in Britain.

Feeding

Generally, donkeys are 'good doers' and do not require the same feeding pattern as horses or ponies, in fact, a donkey that is too well fed is much more likely to encounter health problems. Unlike cattle and sheep, both of which eat large amounts of food at a time and then ruminate afterwards, donkeys have relatively small stomachs and eat only small amounts at a time. When grazing is hard to get at, this may mean that they have to work for their food continually throughout the day, but, in all bar the driest of conditions, they will nearly always manage to find sufficient to keep them going. Having to work for their

Not a case of a good grooming being required, but the Poitou Charente donkey is distinguished by its matted coat and by being extremely rare worldwide – even in his home country of France.

food also has the advantage of preventing them from becoming bored and getting into mischief, although, conversely, if they are too hungry, they may well be making escape attempts in search of pastures new.

In the winter or a very dry summer when no grass at all is growing, they will need some meadow hay in their diet. Take care that it is not mouldy or dusty as this may cause fungal infections and lung problems. New hay is not very digestible and may cause colic so feed only that which is at least six months old.

In really severe weather or as a special treat, donkeys will appreciate a few horse nuts, and unless they are working hard

(unlikely in the smallholder's situation) or in foal, they do not need anything else added to their diet.

RABBITS

We have a farming friend here in France who, like the majority of rural smallholders, keeps several hutches of rabbits for the pot. He has built up quite a reputation for the quality of his stock and sells regularly to outsiders. Despite the ever-increasing number of British people living in this area, almost all our friend's rabbits are sold to the French. Why? Because the ex-pats have the image of a cuddly, fluffy bunny rather than, as the French do, a rabbit being the vital ingredient of a healthy meal.

If, however, you can get away from the idea of rabbits being children's pets, they are well worth considering as an easy meat source. Their housing is minimal and therefore ideal for the small smallholding, and, unlike any of the other meat-producing livestock, rabbits are not subject to any government legislation. The most common breed of rabbit for the table is the New Zealand White, but almost any rabbit type will be suitable provided that it has some size about it – you cannot expect a good meat carcass from small parents.

Housing

Hutches are probably your easiest solution. You can make them yourself or buy them, but, although they are an efficient use of space, it is best not to buy the kind built on two or three levels since, unless you are absolutely scrupulous regarding your daily cleaning routine, some of the urine and spilt water is bound to find its way from the top pens and into the bottom ones.

Movable arks are, in our view, preferable, but you will need to equip them with mesh floors so that, although the inhabitants can still graze at the grass protruding through, they cannot dig their way out. More importantly, the neighbour's dog and predators cannot dig their way in. Such arks also provide protection against the direct weather while still allowing for a good flow of fresh

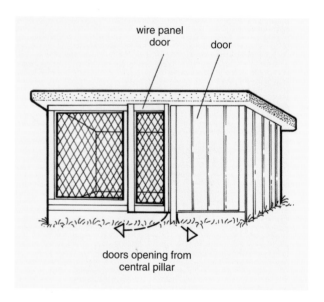

doors opening from
central pillar

*Three styles of
rabbit housing.*

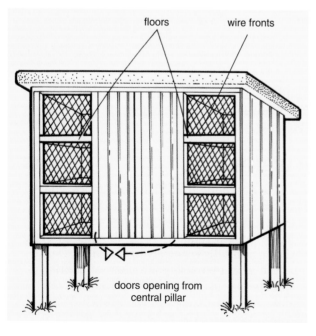

doors opening from
central pillar

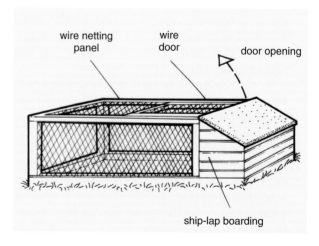

It will be necessary to have several hutches in order to house adult females, youngsters and a resting male.

air; as with all livestock keeping, adequate ventilation is vital for healthy rabbits. An ark should be moved on to fresh ground at least once a day, and possibly two or three times when it contains a doe and her offspring. Movable arks in which it is intended to breed and rear litters should have a portion 'boxed off', and, in order that you can continue moving the run on to fresh ground when the babies are still in their nest, this box should obviously have a solid floor. You will need several hutches or arks when breeding in order to house adult does, youngsters and a resting buck.

Breeding

When a doe is to be mated it is a simple matter of putting her in with the buck (do it the other way around and it is not unusual for the doe to fight with the buck). Females are ready to mate at around five months but they can become pregnant much earlier, so be aware of this possibility when keeping youngsters of both sexes mixed together. Unusually, rabbits do not have to come into season before they can be mated; it is good practice to leave them together for at least 24hr so that you can be reasonably sure that the buck has mated with the doe at least two or three times. A young male in prime condition can service from three to six females a week.

Gestation takes thirty-one days. A day or so before she is due, the doe will start gathering bedding together to make a nest and will also pull out some of the fur from her breast with which to line it. Keep notes of when the pair have mated since in this way you can ensure that the kindling area is cleaned out two or three days before the doe is due to give birth. Do not be tempted into poking about in the nest when the youngsters have been born: a nervous or young doe may well kill her young because of undue disturbance. You will be able to see movement in the nest and, in that way, be able to see that all is well; other than that wait for a few days before investigating,

or, better still, wait until the babies begin to explore after about fourteen days. At eight weeks remove the doe to another hutch or run, thereby leaving the youngsters in familiar surroundings.

Feeding

Young rabbits will wean themselves by nibbling at their mother's solid food. For adult stock ensure that whatever food is used contains a correct balance of protein, fats, fibre, carbohydrates, minerals and vitamins, but remember that some rabbit foods also contain additives such as coccidiostats and probiotics, which are not necessarily good things to be included in the diet of an animal that is eventually going to end up in the

Wait until the young rabbits have left their nest before investigating – disturb them any sooner than this and there is always the risk that the mother will kill her offspring.

human food chain. It is important to have all the different options explained before deciding on what type of food to purchase. Labelling should state whether or not the food is 'complete', 'complementary' or a 'food supplement' – terms that are all self-explanatory. As well as concentrates, rabbits should also be given roughage in the form of hay and succulents in the form of greenstuffs. The hay must be of good quality and the greenstuffs can be any of the natural grasses, garden vegetables and wild plants, such as groundsel, chickweed, dandelions and clover. Be sure that you can accurately identify the wild plants first though: celandines, henbane, poppies and buttercups, for example, are all poisonous to rabbits. It is possible to feed rabbits on just cereals rather than specifically manufactured feedstuffs, but by feeding a proprietary concentrate containing a preventive medication you will help to ensure that your stock is, at the time of its life when it is most vulnerable less likely to succumb to coccidiosis.

Health

Provided that you keep your rabbits well fed, clean and, if using arks, continually moved on to fresh ground in order to lessen the risk of their picking up tape or roundworms, the likelihood of coccidiosis occurring is just about your only worry. This is caused by a small parasite *Eimeria steidae*, and the infected animal usually suffers from diarrhoea, loses weight and huddles in a corner of the hutch. Young rabbits of between six and twelve weeks are particularly susceptible; if such symptoms are noticed, see your vet.

Check on what any newly acquired rabbits have been fed as they do not always take to a sudden change to a new diet and this could, of course, create health problems. Likewise, those used to a drinking bowl may not take readily to the bottle drinker, causing them to become dehydrated. Insufficient water will also cause problems in lactating females who require at least a litre of water a day in order to convert it into milk for her young.

A particularly inquisitive Red Dexter! (Courtesy: Stuart Webb)

ALPACAS

In an effort to enable you to consider all possible livestock and their suitability to your particular smallholding, it is worth just mentioning alpacas (their cousin, the llama, is probably too large for the type of smallholder at whom this book is aimed). As pets they give an enormous amount of enjoyment, but their real value is in their fleece, which is shorn annually, normally in the spring. An animal will produce anywhere between 2 and 5kg (4.5–11lb) of fleece. The Huacaya is the breed most commonly seen, but there is also a second type known as the Suri, whose fleece hangs down, unlike the Huacaya's, which is dense and more like the fleece of a Suffolk sheep.

To find out more about alpacas it is a good idea to buy a copy of *Smallholder* magazine, as there are always breeders advertising there who will be only too pleased to tell you more about any necessary legislation, space and food requirements, breeding bloodlines and general care.

CHAPTER 7

Barns, Buildings, Fields and Fencing

The very small smallholder working in his or her back garden or on a nearby allotment will probably have to try and store everything in a garden shed and keep feedstuffs in a vermin-free dustbin close to the chicken house. Even on a slightly larger set-up it will probably not be possible to have too many buildings and outhouses; not least because of any planning restrictions that the local council might impose. Nevertheless, the fact remains that at least minimal storage and housing will be required – it might be rather grand to call a small, corrugated lean-to 'The Barn', but

some form of place may well be required for keeping hay and bedding dry, so, if it is a barn to you, what is wrong with that?

Much depends on the degree of self-sufficiency and smallholding practice that you are considering: for the storing of tools to tend a few vegetables a garden shed is all that is required. It needs a strong concrete base and, if it is intended to use a part of it as a place for storing harvested crops over winter, it must be situated somewhere where there is the least likelihood of frost. A home for chickens and poultry will, on the smallest of places,

Unfortunately, not many smallholders can aspire to a barn of this grandeur; note the incorporated 'pigeon loft' above the doors.

A movable house and run is always the best option for a few chickens or bantams. (Courtesy: Rupert Stephenson)

probably be a combined house and run, which, it has to be said, has many advantages over the traditional poultry pen. A movable house and run, ark, fold unit – call it what you will – is likely to be warmer; better protected from the elements, allow the inhabitants semi-freedom (and they can always be allowed out for a potter round the garden when someone is on hand to see to them) and provide security from predators. Perhaps most importantly, however, the whole thing can be moved on to fresh ground from time to time, thus ensuring that there is a continual supply of greenstuffs, insects and less chance of a build-up of disease.

It is unlikely that any existing buildings on your smallholding will, unless they have been specifically built for such a purpose, be absolutely ideal. For the majority, therefore, it will probably be a case of make do and mend;

but, come what may, the basic requirements of all housing is that it should be dry and free from draughts and yet have an adequate air supply. Many old properties have a pig sty – a legacy of the days when every householder used to keep and butcher his own animals. The traditional stone sty with a yard in front evolved through centuries of trial and error and, according to many pig owners, has never been bettered; if there happens to still be one on your property, you are already ready to start. Pigs seem to like this particular method of housing, provided that there is plenty of bedding in the sleeping quarters, the system is a good way of allowing the inhabitants access to fresh air without letting them root up pasture land. Other buildings may be equally suitable, but the insulation must be good, with no condensation.

Those in the fortunate position of being able to consider the purchase of somewhat larger livestock could do no better than create a small 'yard' of loose boxes of a size that would house a small Dexter cow, a donkey, in-lamb ewes or, split into sections and stalls, goats, poultry, or even a hutch or two of rabbits. In many cases, it will also prove useful if the building or outhouse permits free access to and from the field or paddock – in winter, some animals will prefer to spend time in the shed rather than standing knee-high in mud, while in summer they should still be allowed access to the building at all times so that they have shelter from the sun and flies.

BUILDING REQUIREMENTS

Before considering the erection of new build-ings or altering existing ones, it makes sense to contact your local council and find out exactly what is and is not allowed under their particular bye-laws, public health legislation and planning permission requirements. Generally, the best animal housing is con-structed from brick, stone or concrete and lined with wood as insulation, but timber, prefabricated, stable-type buildings are warm and serviceable. Where possible, avoid using buildings that are predominantly made from corrugated metal as these are cold in winter and unbearably hot during the summer months.

Roof, Floor and Doors

The roof should be made either of wood over which has been laid good quality roofing felt (use the thickest grade you can find as, with care, it will last a lifetime, unlike the cheaper

A large building or loose box can be usefully split into sections and stalls to accommodate goats and sheep.

A decent roof overhang and some guttering will help to prevent the area immediately surrounding the building from becoming a quagmire.

Roofing Felt Tip

When working with roofing felt, make sure that it coincides with a warm, sunny day and roll lengths out on the ground for a while in order that the sun may soften the material and makes it more pliable. It is far easier to use and a better finished job is achieved. Also, consider using lengths of batten to seal the strips rather than to just bang in a multitude of felt nails, each of which will pierce the felt and make it potentially weaker at that point.

products which are nothing more than tarred paper) or slates and tiles (unfortunately, a much more expensive option and one that requires a certain amount of expertise if one is to be constructed properly). There are now 'corrugated' roofing options other than metal or asbestos and these too will last, are relatively cheap and also have the advantage of offering good insulating properties. Beware of using feather-edge boarding as a roof; while it makes adequate side walls, especially if double-skinned, it is never going to be completely waterproof as a roofing material. You could, of course, use it and then cover it with roofing felt.

In height, a span building for livestock should be about 2.5m (8ft) at the apex and a lean-to around 9m (9ft 6in) at the front, sloping to 2m (6ft 6in) at the rear. Not only will a building of this height accommodate a donkey or a house cow, but it will make life easier for yourself, as well as providing space for a good air flow and that all-important ventilation. A stuffy building is likely to foster colds and other problems, especially when the conditions thus created are mixed with the inevitable smell of ammonia from urine.

Floors

A concrete floor is probably the cheapest to construct and the easiest to keep clean. It should be higher than the ground outside to prevent flooding and, preferably, have a slope towards a drainage point. It must also be of sufficient thickness and strength to avoid the possibility of its becoming cracked and breaking up. Make sure that it has been laid on a thick base of rubble – 23cm (9in) would not be too much – and that the rubble has been completely compacted before the concrete is laid. In order to further aid drainage, create a herring-bone pattern in the concrete before it has completely dried – this can be done with the edge of a piece of timber.

A building which is never intended to house anything more substantial than a flock of free-range chickens, for example, obviously does not need such a heavy floor (nor the height and size dimensions laid out above)

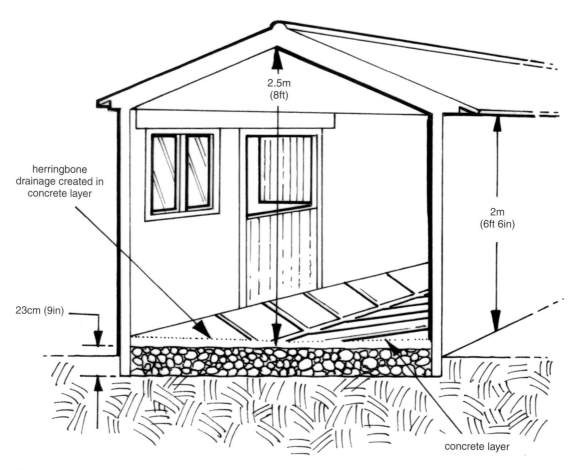

2.5m
(8ft)

herringbone
drainage created in
concrete layer

2m
(6ft 6in)

23cm (9in)

concrete layer

Height, ventilation and floor drainage are all important considerations when it comes to constructing a building for livestock.

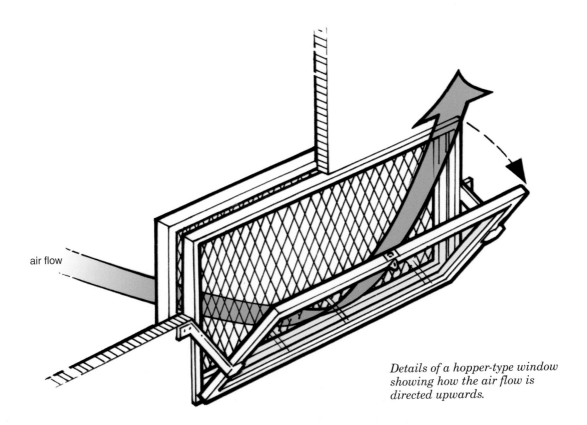

air flow

Details of a hopper-type window showing how the air flow is directed upwards.

and wood may be the preferred alternative. Ensure that it is not resting on the ground, which will encourage premature rotting and also vermin, and also that there are sufficient cross-joists so that the floor does not 'spring' every time you enter.

Doors and Windows

Fit half-doors where possible and practicable, they help in providing ventilation and the animal inhabitants will enjoy being able to look over the door and see what is going on – all animals are inquisitive and curious and a view on the day-to-day world will help them from becoming bored. Cold weather affects most animals less than does a draughty environment and so it should be possible to keep a half-door open in all but the most severe of winter conditions. Goats are something of an exception in that they can withstand heat better than cold or wet, so they must be kept in during winter nights or during inclement weather.

Windows should also be capable of being opened, and, in an ideal world, should be of the 'hopper' type, hinged from the bottom and opening inwards; air entering the building in this way is directed upwards and, being cooler and therefore heavier than the air already inside, will fall slowly and, in doing so, will become diffused and warmed. In some situations you may wish to screen the windows with some kind of mesh in order to prevent access by vermin.

Fixtures and Fittings

Exactly what fixtures and fittings are required will, of course, depend entirely on what is being housed. All poultry, with the exception of ducks and geese, for example, will need some form of perching and the inclusion of nest boxes.

Some chicken houses have a nest box fitted to the outside wall for easy accessibility.

Perches and Nest Boxes

Make the height of the perches appropriate to the type of chicken you want to keep, but, as a general rule, they should be about 60cm (2ft) from the ground and not less than 25cm (10in) away from the wall. Desirably, the perches should be made of planed wood so that they are easy to disinfect and less likely to splinter. The tops should be rounded and about 5cm (2in) wide. Allow at least 20cm (8in) of perch space per bird. Install nest boxes in the darkest part of the building – usually this is directly underneath the windows – and away from the perches so that the birds are less inclined to roost in them. Nest boxes may be constructed, perhaps in tiers or as a row of three or four, with each box being closed in on three sides for privacy.

Allow one box for every three hens, and, for large fowl, box dimensions of around 60cm (2ft) square.

Lighting

Never forget the winter months; once you begin smallholding you will soon realize that there are more of them than there are summer ones. No matter what livestock is being kept (and even if the building is only to be used as storage), some form of lighting is always handy and frequently essential. It may be that, in some situations where your buildings are away from the house or a convenient electricity source, you will have no alternative but to incorporate the price of a generator into your initial budget costs. Bearing in mind the outlay, it is important to buy a generator that is suitable for the job you have in mind. A petrol one is generally cheaper and quieter, while diesel has the advantage of being economical in prolonged use (and most models have a larger fuel tank capacity than do petrol ones).

Laying poultry will continue to lay throughout the winter months if they have a regular amount of 'daylight' hours during which they can feed, drink and exercise. Obviously, in the depths of winter, this can only be achieved by some form of artificial lighting. A light source can be advantageous to other livestock too, but not necessarily for their benefit – more for the person charged with feeding and mucking-out in the dark before going to or returning from work or school.

If you decide to supply power via an extension from your own home or from another building, installation should be carried out only by a professional electrician. Not only is there the obvious danger aspect to consider but also the fact that, since 2005, although some DIY electrical work is still allowed, it must comply with the requirements of Part P of the Building Regulations. Again for reasons of safety, do not locate any lights within reaching distance of any livestock being kept there, and remember that a goat can be quite tall when on its hind legs – a stance they seem to quite favour. Always cover light bulbs with a secure wire cage: some lighting units are, in fact, supplied with them as a matter of course, but, if not, it is possible to buy one at most agricultural or electrical suppliers.

Feeders and Drinkers

Housing for cattle, goats and sheep should, for preference, include a wall-mounted hay

An old hay turner, adapted for use as an outdoor hay rack.

Hay Nets Warning

Hay nets are useful, but it is absolutely essential that they are tied securely and at a height whereby it is impossible for an animal's leg to become caught within the mesh and suffer subsequent serious injury.

rack and a water trough. This should be an automatic one that the animal operates by pressing a plate with its nose. A feed trough should also be included and must be securely attached to the wall. Unless it is specifically designed for livestock, a movable container tends to spend more time upturned than upright, although using a bucket as a feeder and wedging it into an old car tyre usually

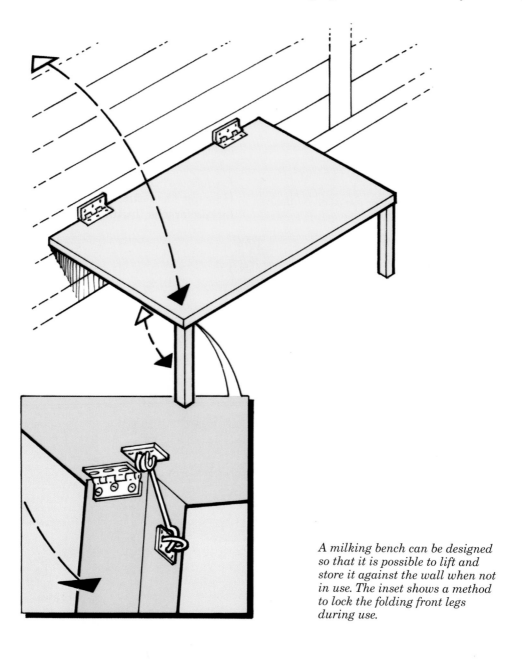

A milking bench can be designed so that it is possible to lift and store it against the wall when not in use. The inset shows a method to lock the folding front legs during use.

solves the problem. There is one school of thought, but not one to which we subscribe, that believes fittings such as a manger and hay racks inside the building are best kept to a minimum. The argument for their being placed outside is that, by doing so, valuable internal space is freed and also that the smaller the number of projections, the smaller the possibility of injury.

In the interests of hygiene, one thing that is important is the need for feeders and drinkers to be removable. Obviously this is not possible with automatic drinkers, but it should still be a matter of priority to see that they are kept free of hay seed and errant food particles. Galvanized or alloy utensils are more robust than even heavy-duty plastic, but they are, unfortunately, often as much as twice the price. Both will clean just as easily, although it has to be said that the plastic probably has the edge as they are usually constructed in a mould, whereas metallic feeders and drinkers have seams and lips which are sometimes difficult to scrub. Under no circumstances be tempted into making your own feeders out of wood – they may have a certain charming nativity-scene naivety about them, but they will be extremely irksome to clean out thoroughly. There is also the very real chance that second-hand timber may have been treated with paint or a preservative harmful to livestock.

Milking Benches

If you are considering goat-keeping, a milking bench is a useful accessory as a goat standing on the ground is too low to be milked comfortably. A bench raises her by around 30cm (1ft) so you can sit on the edge of the bench or on a stool while milking. Made of wood and around 60cm (2ft) wide, a bench can be hinged to a wall at one end with two strong legs (also on hinges) at the other end, so that it can be folded flat against the wall when not in use. Most benches you will see have a point at which a feed bucket can be attached, as the majority of handlers feed their animals while milking. There is a counter-argument to this

which states that goats should not be fed during milking as it makes them impatient to get at the food and more likely to end up with a hind foot in the milking pail.

But even if she is willing to make the attempt, it could be a different matter trying to lift a cow physically on to a milking bench. You will, however, need some place in which to milk your cow, where she can be tied loosely and which is relatively clean. Legislation requires that cow's milk for drinking is pasteurized, that the 'dairy' and the 'cooling room' meet certain standards, especially if you are offering milk for sale to the public – in which case you will also be subject to periodic inspection.

Food Stores

The positioning and type of building most suitable as a food store is well worth a mention here. Unless you are prepared to waste produce and money, a leaking lean-to will not do. The shed should be off the ground and in a position where air can circulate all around. It should be dry and, most importantly, totally vermin-proof. You can help to ensure against vermin by keeping a regular check on likely entrances and exits, and also a constant supply of rat poison both in the building and in safe strategic baiting points outside. Check all baiting points regularly, infrequent baiting is as useless as doing no baiting at all and may lead to a build-up of immunity. Make sure that they are inaccessible to your stock or any domestic animals. For that reason alone, it may help to use the types of bait that are impossible for rats and mice to carry off, only to leave them in vulnerable places where they could, for example, be pecked at by poultry or gnawed at by inquisitive ruminants. Internally, all food, whether it be bagged or baled, should be stored off the ground – even a concrete floor is never completely dry so it is necessary to ensure that air can circulate freely (as a guide to the size of the building required for hay or straw, one tonne of baled hay – about 40 bales – takes up approximately 84cu.ft).

Metal, not plastic, dustbins make the best feed storage containers. (Courtesy: Rupert Stephenson)

Where space precludes the provision of a specific building or store at the very least one should keep food in vermin-proof storage bins. They might be only galvanized dustbins (do not be tempted into using plastic, as rats and squirrels will soon make short work of gnawing through at the most vulnerable points), but far better are the old-fashioned, but nevertheless serviceable, sloping-lidded feed bins, much favoured by the horsemen and poultry fanciers of bygone years.

Building Security

An isolated property, although perfect for the purposes of a smallholding, can be an attraction to burglars, thieves and vandals. The simplest and most effective method of minimizing the risks is to have your animals kept as close to the house as possible and to secure doors and gates with a decent padlock at all times. An outdoor light that switches itself on as soon as a beam is broken is another deterrent, although the beam must be arranged at

Store Food Efficiently

Buying in a large stock of food may seem economical and cost effective, but only if it can be properly stored on wooden pallets and away from the walls so that air can circulate. Always keep a note of what food is stored where and try to develop a rotation system whereby the first food purchased is the first to be used – it is all too easy to store food next to the door and use it from this, the most obviously accessible pile. If you do this, the older food will go stale and unpalatable and any preventive drugs, medicines and vitamins will lose their effectiveness once they are out of date. Get into the habit of checking the sell-by dates on purchase and never use food that has gone beyond this time.

Vehicle Security

While on the subject of security, it is as well for the smallholder to remember that, in addition to the barns and buildings themselves, some of their contents should be separately secured. Remember to put some kind of lock on livestock transporters or any other trailers used about the smallholding, and be especially vigilant if you keep a quad bike, as these are particularly vulnerable to theft. Be sure that you are honest in telling your vehicle insurer where it is kept – an out-of-the-way place might mean that you have to pay a higher premium – but better that than finding out that your insurance will not cover you in the event of the trailer or cycle being stolen.

Any vehicle of use on the smallholding is particularly vulnerable to theft and should be kept securely under lock and key.

such a height that it is broken only by human movement and not that of the shed's inhabitants or passing dogs or cats.

Any workshops or tool sheds are an obvious target, and these must also be kept locked and any windows shuttered with something better than just a sheet of corrugated metal propped closed by a couple of tree branches. Being tucked away out of sight and sound of the main house, will give the thief all the time in the world to force an entry. An alarm system will give extra protection and is not too expensive to purchase and install.

FIELDS

How wonderful it would be if every smallholding could contain an orchard. If it did, it would, provided that the trees were protected from the unwanted attention of livestock kept in it, prove the ideal place for all types of poultry and animals. The trees would offer shelter from the elements and the ground at their base grazing for all.

Every smallholder who keeps livestock will need a little grassland of some description. It may only be a small area of reasonably level and closely-cropped sward on which to place a coop or two of hens and chicks, or it may be a paddock split into grazing plots for sheep or goats. Any land you have available benefits from being split into four plots which can be eaten off in rotation. Two of the areas could perhaps be used for a hay crop, but, on a small acreage of ground where grass is likely to be at a premium, it might be better to buy hay from a neighbour and leave yourself with the extra grazing area. A field, even when

Any available grazing will benefit from the protection of a hedge or tree line.

Newly planted trees need protection from livestock, and even where close-croppers such as sheep are being grazed, grassland will require 'topping' at least once a year in order to prevent pasture land from becoming rough and unkempt.

regularly grazed, will require to be topped at least once every summer – all stock prefer short, freshly grown grass. Unfortunately, animals eat some vegetation in preference to others and anything not eaten grows rough and unkempt. Horses and donkeys are the worst offenders and will not eat in areas contaminated by their own droppings. Thus if you decide to keep either, it is therefore important that you remove any manure on a daily basis: if you stack it near to your vegetable garden, it will prove a useful addition to your autumn digging.

Fertilizing

Soil is able to supply some or all of the nutrients for grass growth, but fertilizers help to make up the difference between what is there naturally and what a particular crop needs to

Ragwort

The banes of any smallholder's life are docks, thistles, nettles and, in particular, yellow ragwort (which is poisonous to stock, whether it is growing, cut or dead). Take every opportunity to eradicate them all, but especially the last, which is seemingly increasing on an annual basis. Pulling ragwort by hand – wearing a strong pair of garden gloves – will, in eight cases out of ten, remove the plant roots and all. Docks and nettles, being far more deeply rooted, will prove more difficult and necessitate some serious and careful digging. If the nuisance from weeds on your land eventually affects others you may find yourself served with a notice from the local authority. On registered land, a notice may come from DEFRA

Without careful maintenance, paddocks soon become overgrown with weeds.

grow. Even though grazing animals recirculate a high proportion of nitrogen, it may still pay to give your paddocks a good dressing of nitrogenous fertilizer during the early growing months of the year, that is, in late spring. If they are small plots you could try spreading it by hand – it comes in granular form and it is a simple operation to scatter it across a field from a bucket. Make sure that you wear rubber gloves, however. Although fertilizers are not normally harmful, they can cause cut or damaged skin to sting. Grazing land may also need a dressing of lime every three or four years. A visit from a soil analyst to determine exactly what the soil requires is to be recommended. Those particularly concerned

with the subject of organic smallholding will find such a visit essential.

Weed Control

Control weeds in grassland with selective weedkillers. Better still, cut them and then, when dry, rake them up and burn them. Well, that is the theory, but, in reality, to let weeds dry and then rake them into a pile suitable for burning, will only encourage weed seeds to drop from what is being raked and subsequently spread over a greater distance in future years. If this method is, however, your preferred choice, make sure that grassland weeds are cut before being allowed to flower. Despite the ideals of organic care, spot treatment with a weedkiller is sometimes the only way to control persistent weeds. Likewise, brambles and other woody-stemmed weeds can be killed only with a brushwood killer, unless you are prepared to try a flame gun (popular twenty to thirty years ago, but less so now). The true organic evangelists maintain that weeds should neither be burned nor killed: they recommend that they should be hoed out and left to die. This will, however, work only when the surrounding grassland is very short (or newly seeded), the weeds are very young and when the weather is dry enough to shrivel the young plants immediately after being hoed. Such methods are useless in wet weather, as the weeds merely lie on the surface and send down new shoots.

FENCING

What is meant by fencing? A dictionary would variously define it as a defining and/or protective border, and so it is, but to the smallholder it is much more besides. It is also, in almost all its guises, expensive. The only exception might be when it comes to poultry fencing made of wire-netting or flexi-net, which will keep sheep and goats where they should be by the use of an electric fence energizer.

Some pig breeds, such as the Large Black are extremely docile, a factor that makes them well worth considering when it comes to the type of fencing necessary; this placid temperament enables them to be easily contained by a single strand electric fence. Even a back garden vegetable patch which borders a stretch of farmland may need protection from neighbouring rabbits. But whatever, the cost of any of this type of fencing must be mentioned in pounds rather than pence. If money were no problem every fence should be of the post-and-rail type – it is good to see and the treated timber will last for several decades.

Boundary Fencing

Your boundary fence is the most important one and must be properly maintained. If your internal fencing fails and sheep or goats eat

A good electric fence energizer is essential when using single-strand or flexi-net fencing.

Post and rail fencing is wonderful to see, but can be prohibitively expensive.

Barbed Wire

Barbed wire is a mixed blessing on the smallholding. There are some who think that Joseph Glidden, who invented it in 1873, should be compelled to lie on a bed of it for all eternity. Certainly it should be treated with care and consideration: using it as described above will almost guarantee that stock will stay where you want them and that an unwanted human presence will be made more difficult. On the other hand, there are many unfortunate proven instances where a dog or a wild deer has become trapped between two strands and suffered greatly as a result.

all your precious vegetables, it is only you who suffers. If your boundary fence gives way you may have to pacify and possibly compensate an irate neighbour. By far the most effective boundary fencing is stock fence (sometimes known as pig-netting). The mesh should be clipped with wire ties to strained plain wire that has been fixed at the top, the middle and the bottom to pressure-treated timber posts, firmly fixed into the ground every few metres. The posts should be long enough to protrude above the fencing by at least 30cm (1ft), as this will be required to carry two strands of barbed wire.

Hedging as Boundary Fencing
Although the plants (or 'whips') will need to be protected by some form of wire fencing as

they become established, a hedge around the boundary is beneficial in many ways. In the small garden it acts as a windbreak to the vegetable plot and prevents your activities from being seen by the neighbours. It will also, in a small fashion, help in protecting the sound of foraging chickens, for example, from being carried to next door. On a larger smallholding a mature hedge can not only mark your boundary, but will also give shelter to livestock, filter noise and dust, and possibly help to prevent the spread of some airborne diseases. In addition, a good hedge that contains holly, hawthorn and other traditional (non-poisonous) species will go some way towards protecting your property from unwanted visitors and will, of course, do what is after all, a hedge's main purpose, keep your livestock where they belong. But as with all things, there are some disadvantages: a thick, unruly hedge will shade out grass and plants struggling to grow at its base and spread its roots into the field or garden; thereby impeding rotovating or digging activities and requiring regular annual maintenance.

Although the newly planted hedge needs some protection in the early stages, a mature boundary hedge will eventually be beneficial in many ways.

Electric Fencing

Probably the cheapest of all options, single-strand, electric fencing has its uses – it will prevent horses and cattle from leaning too heavily on an existing stock fence and also as a temporary barrier preventing livestock from gorging on seasonal fruits, some of which, such as acorns, will prove harmful, and even fatal, to certain species, especially equines (pigs, strangely enough, thrive on the fruits of oak trees). Flexi-net electric netting will, as indicated earlier, prove to be a useful way of containing sheep and goats and also of deterring foxes from attacking a paddock of free-range poultry. The bottom of the wire is not electrified in order to prevent an electric short to the soil so it is therefore important to keep vegetation clear of the next strand.

Wire Netting

Wire netting is useful in the garden, it will, as mentioned earlier, prevent the vegetable patch from being decimated by local rabbits. It is essential when constructing a chicken run, but surplus rolls of it have a multitude of other uses. Laid over flower beds in the winter when vegetation is low, it can prevent free-range chickens from irredeemably

Livestock Fencing

Your property may require specialized fencing. Poultry need protection from predators more than they need to be prevented from escaping from a particular situation, whereas goats, for instance, need to be protected from themselves – being of an inquisitive nature, they know that the grass (or bush or tree) is always greener on the other side of the fence. To prevent them from climbing out of their enclosure, use stock fencing raised around 15cm (6in) off the ground with an extra (single strand or barbed) wire fixed at the same distance above the top of the fence. The corner posts must obviously have wire rather than wooden braces, which would allow an active goat to climb to freedom.

damaging the roots and shoots of precious flowers and shrubs. Larger mesh is commonly used for poultry fencing and also as a climbing frame for the likes of runner beans in the vegetable patch. Small-mesh, heavy-density netting can, quite profitably, be utilized stretched under coops and runs containing a hen and her chicks, and doing so will prevent the chicks from dusting their way out and rats from scratching their way in; while still allowing 'mum' and her babies access to fresh grass protruding through the mesh.

Drystone Walls

While not technically fencing, drystone walls have, nevertheless, in certain parts of the

Wire netting has a variety of uses on the smallholding, but is most commonly used with poultry. However, the height here would not be sufficient to prevent foxes and straying dogs from entering and its success is dependent on the additional electric perimeter fence, the strands of which can just be seen.

country been a traditional method of containing livestock. Originally, such 'fencing' was, although undoubtedly useful in creating enclosures, also a means whereby landlords kept their outdoor staff employed during the quiet winter months. They are now an easily identifiable part of the British scene, but, with the lack of ready labour, any breaks in the walls have, over the past fifty years or so, probably been replaced by posts and strands of barbed wire rather than being rebuilt properly. If your chosen smallholding happens to include such topography it is beholden on you to maintain the walls in the best ways possible. Drystone walling is undoubtedly an art and, unfortunately, those skilled craftsmen who used to repair such things as a matter of

course are slowly dying out, although there are instructional courses available which the novice wall builder will find extremely useful. In essence, the construction of such a wall involves knowing that the base should extend well beyond its apex in width, that either side should be constructed of solid stone and the inner space from shingle-sized, otherwise unusable material. At intervals throughout the length of the wall 'through ties' should be incorporated.

Drystone walls are, in certain parts of the country, a traditional method of containing livestock. There is, however, a great deal of skill involved in their maintenance and repair.

CHAPTER 8

Waste Nothing, Enjoy it All

One only has to listen to the likes of Bob Flowerdew (bearing in mind his subsequent chosen career, never has a man been so aptly named) on BBC Radio 4's 'Gardeners' Question-time', to know that, with a little forethought and ingenuity, the smallholder has much scope to utilize what others would waste. Composting is an obvious way of ensuring that nothing is wasted and the whole point of the enterprise is that the quite literal fruits of hard work will, in season, be enjoyed. There are, however, other slightly more subtle definitions meant by the title of this chapter. These may include cost-cutting exercises, recycling, the harvesting of hedgerow plants in order to make jams, alcoholic drinks and, possibly, when the smallholding bug really takes hold, the showing and exhibiting of one's own products and livestock.

To get the best from being a smallholder in a new area one should make every effort to integrate at whatever level is felt to be appropriate. Although village life is not quite as portrayed on television's 'The Vicar of Dibley', it can feel not dissimilar on occasions. Everything you do around the smallholding will be watched and commented upon – probably with suspicion – for quite some time after your arrival. Allay these fears by joining as many country- and farming-associated local clubs and groups as you feel able, bearing in mind the restraints of work, time and money (beware, however, of becoming the village equivalent of Linda Snell on Radio 4's 'The Archers'). Even the more urban smallholder should join a group of people whose activities interest him or her most – in both town and country there are garden groups, self-sufficiency organizations, poultry clubs and farming-affiliated associations, all of which hold regular meetings, have trips out, competitions, seasonal shows and, perhaps most importantly of all, offer the chance to meet experienced gardeners, fanciers and smallholders. It would be a rare thing if none of them were prepared to help or advise the less-knowledgeable, but equally as enthusiastic members, whether it be with gifts of surplus potato tubers or advice on how to label a jar of home-made jam in preparation for the autumn show.

RECYCLING

There are many ways of cutting costs without cutting corners. Growbags, for example, have their place in the self-sufficiency garden and, provided that they do not contain any disease spores or chafer beetle larvae at the end of a season, the soil contents can be added either to the raised beds or the compost heap once they are finished with. The smallholder can do worse than collect newspapers from friends and family and use them as a bed for animals of all sorts. Of course, it is possible to buy bales of recycled and shredded paper from your local agricultural supplier, but the whole point is to recycle and save money. It is also possible to use dried leaves as animal

It pays to integrate into village life, whether as a member of the parish council, smallholding club or simply by participating in the family dog show at the village fete.

bedding in areas where they are plentiful and they fulfil the ideals of being both recycled and money-saving. Their only slight disadvantage is in the fact that they are bulky to store until required, but, once again, they provide a perfect addition to the compost heap; in that way they could be said to be 'multi-cycled': first from the tree, then as bedding before being composted and, lastly, as a growing medium – a system that should appeal to all smallholders.

Keep It Tidy

Beware of becoming too untidy in your desires to recycle and save money – carpets, old freezers and tyres might have their place on the smallholding as mulching mats, feed bins and potato towers, but the look of them is not likely to endear you to either the neighbours or the local authorities.

Fishmongers and your local fish and chip shop take delivery of their fish in polystyrene boxes which are non-returnable. They might smell a little, but, if you can beg some, they make excellent seed trays due to the facts that they are deep and naturally insulated.

Composting

The most obvious and beneficial form of recycling around the smallholding has to be in the cultivation of a compost heap, but it must be done correctly. So many people think that a rotting pile of vegetable matter left to its own devices in the corner will, in a few weeks' time, magically transform itself into a prime-potting medium. Unfortunately, that is rarely the case and the very minimum required is a three-sided structure complete with a moisture-retaining base. The dimensions should not be too large as the whole point is to build up to a reasonable height (perhaps around a metre or so) so that the many bacteria and enzymes can get to work in the warmth and moisture thus created. Mixed with bedding from the livestock and regularly turned so that the bottom layer becomes the top, you should be able to see some positive results in about six months. Covering the top of the heap with an old carpet or thick wad of straw will also help in accelerating the process and plants grown in the resultant mixture will taste all the better for having been cultivated in 'free' compost.

Any material that was once alive can be added, but beware of including too many lawn clippings as these will just turn into a hot, soggy mess without the proper care and attention. Woody articles should be avoided as they will obviously take longer to break-down than any of the softer stuff like cabbage leaves (for more details on composting *see* Chapter 2).

Newspapers and Sacks
Roll up strips of newspapers into toilet-roll sized tubes and staple them to form planting pots for vegetables. Filled with soil (free from the heap, of course), bean seeds and the like can be individually planted and, once large enough, planted out, paper and all. Provided that the amount of paper used to make the 'pot' is not too great, the roots of the plant will force their way through, while the tube should stay firm enough not to disintegrate until planted in the vegetable patch.

A compost bin should be a rigid-sided structure complete with a moisture-retaining lining.

Be wary of giving water to poultry in troughs to which wild birds have access as it is possible that their faeces may cause it to become contaminated.

A neighbour here in France lines the bottom of his runner bean trench with newspapers before topping up with farmyard compost. The two combined certainly seem to help to retain moisture around the plant roots when all about is wilting and in need of nightly watering from the well. Likewise, newspaper is wonderful when used as animal bedding once it has been torn into strips, as it is absorbent, extremely warm and, most importantly, unlikely to be infested with lice and fleas. Of course, it would not be practicable to shred sufficient newspaper for bedding for large animals (although it is possible to buy a large shredder for the purpose), but used in smaller amounts, it is excellent for dog kennels, rabbit hutches or poultry nest boxes. Once it has been used for this purpose, small quantities may be periodically added to the compost heap.

Save and store all the paper sacks that contained feed for your chickens and livestock; yes, we know that they make a good base for a bonfire, but, more usefully, they provide ideal storage for potatoes and similar crops. Recycle nets that previously contained horse carrots in the same way, and, if you find a source of hessian sacks, guard it with your life since they are so useful and hard to come by these days.

Water
As a precaution against water shortages, spring water may be stored in tanks or

concrete containers. Rainwater is often collected for use in the garden, but it is also possible to filter and reuse it in the home for everything except human consumption. A suitable filter system is reliant on adequate rainfall levels and may be expensive to install, but it can often supply between half and three-quarters of all household water requirements and is therefore worthwhile considering.

To recycle rainwater for the garden or even as a water supply for livestock, it is a simple

matter of collecting it from the roofs via guttering and storing it in suitable containers to which a conventional tap has been fixed. Always remember to place any tanks or containers at such a height that it is possible to place and fill a bucket from under the tap. Also, be wary of giving poultry water that has been collected from a roof to which wild birds have access, as it is just possible that their faeces will pollute the stored water – a factor that is especially important with the ever-present possibility of avian influenza.

Some Final Ideas …

Other forms of recycling might not be so obvious. In the past, during particularly cold spring weather, a modicum of success has

Vegetable crops may need overhead protection from pigeons; here, wire netting fencing has been installed to prevent the unwanted attention of rabbits.

been achieved by cutting out the bottoms of plastic plant pots in which tree saplings had been grown (stored in the shed in the hope that one day they might prove useful) and placing them as collars around young cabbage plants. They kept the spring winds off, funnelled water to the roots direct and seemed useful in preventing attacks by insects such as flea beetle.

Fencing stakes that have rotted at the bottom and need to be replaced could be sawn up for winter fuel, but equally, the top two-thirds could be saved and used again as corner struts in a new fence. Chestnut fencing is especially good for this as although the points may have rotted, the main shank will remain sound for many years.

Hosepipe through which straining wire has been threaded makes for good tree ties and prevents the bark of young trees from becoming chafed in the breeze.

Calor-gas-operated bird-scarers are expensive to buy, but fertilizer sacks hung in trees to twist in the wind will be just as effective; it is, however, necessary to be aware of their breaking from their string and ending up as an untidy mess in the ditch. Discarded compact discs might keep birds away from the small garden, but it would take the entire stock of Our Price Records to deter pigeons from the average smallholder's vegetable plot. Video or cassette tape, strung tightly between fence posts or tied to the handles of 25ltr drums weighted with water, 'hums' effectively

Any hazel growing on the smallholding can be cut in the winter for use as pea or bean sticks and, in the autumn, may also supply a harvest of cobnuts.

in even the slightest breeze and has the desired effect of unsettling avian marauders.

Save Money

Apart from helping to save the environment and money by recycling, some lateral thinking can do both. In a glass-to-ground greenhouse that is used all the year round many people conserve heat by lining it with bubble wrap-type insulation. All well and good, but the authors have noticed on several occasions where the bubble wrap has been fixed only as far down as the staging, with the result that much valuable warm air has been lost through the uninsulated glass below the staging. When this error has been rectified the saving on heat has proved substantial.

During the Second World War it was a common practice to cut out each eye of a potato as it germinated and plant it in a small pot of compost. Each seed potato produced, over the weeks of late winter/early spring, over a dozen eyes – having being deprived of one, it simply grew another. After being kept in a large garden frame until all danger of frost had gone, the potato 'seedlings' were planted in the vegetable patch where they grew as well as if not better than those planted in the conventional way. By leaving the side shoots on outdoor tomatoes these can eventually be taken as cuttings and then rooted in pots in the greenhouse in order to extend the season.

Save money on pea sticks and bean sticks by cutting your own. If the smallholding has a little woodland containing a few hazel stems you can coppice these for use in the vegetable garden, and, by doing so. you will also strengthen the hazel bush itself by forcing it to produce new, stronger stems. In a couple of years these should, in turn, be ready to harvest for the next batch of poles and sticks. The cutting needs to be done in the autumn or the winter and before the sap rises. At the end of the growing season it is probably best to burn hazel sticks and stakes as they may well be covered in unwanted fungus spores.

HARVEST FROM THE WILD

In rural areas of France even today it is quite common to come across a local and his wife scything at some particularly luscious looking verge grass, or with their rear ends upwards as they reach into the innermost recesses of a ditch in search of wild herbs. With almost every householder being, by definition, a smallholder, they augment the grass on their land by picking what they can for free with the intention of feeding it to their goats and rabbits. There is a deliberate policy that the councils will not cut their verges until the seeds of its grasses, flowers and herbs have set, so as to ensure that all will propagate again in forthcoming seasons. In addition, many country lanes are virtually car-free and, since in most parts of France there is no need to salt the roads in winter, the result is that the verges are not detrimentally affected by either this or car fumes. It is, therefore, the perfect foraging ground for herbage with which to feed livestock and is a practice that can be emulated in the more rural parts of Britain. Remember always to leave enough of everything in order to ensure that it can regenerate, and check too that you are not breaking the law by picking certain species.

A bit of foraging on the wild side can produce some interesting results for humans too. We all like to think that we are getting something for nothing, but there is also a sentiment that, if we have not paid for it, it cannot be much good. Never has this been less true than when considering the natural crops and harvests of the countryside.

Mushrooms, Herbs, Weeds, Nuts and Berries

Some types of fungus will grow in only certain types of soil; morels, for example, prefer the edges of broad-leaved woodland, whereas the easily recognizable field mushrooms are found in open fields and meadows – and, by doing so, they confound the theory that mushrooms grow well only in damp, shady places. The different types of fungus

Some hedgerow foraging can produce an interesting harvest, the results of which will taste all the better for being free.

come into season at different times of the year, but it is a traditional autumn treat to be out and about in the early morning. Only bother to look in places where chemical fertilizers and sprays have not been used since most types of spore will not grow where such treatment has been carried out. Carry a basket rather than a plastic bag in order to prevent damage to the mushrooms and, most importantly, make sure that you know without doubt exactly what you are picking. A good field guide containing excellent photographs rather than line illustrations is essential.

Herbs and Weeds

As a source of wild herbs, the countryside cannot be bettered. If you know what you are looking for and the time of year at which you are most likely to find a particular herb; it should be possible to pick wild garlic, chives, fennel, horseradish and chervil (cow parsley) in most parts of the British Isles. Some are definitely seasonable, the wild garlic, for example, seen growing in many places, but particularly on the roadside banks of Cornwall, is available only during April–May time, whereas wild chive can be found almost all the year round. Nettles, although a nuisance in the wrong place, are an infallible sign of rich fertile soil. On the plus side, they take in the nitrogen oxides from traffic fumes and give back healthy air. In addition, young nettles make an excellent substitute for spinach and can be used as an accompaniment to many dishes.

Nuts and Berries

Hazelnuts are common enough to find in the summer as they are forming, but not so easy when they are ripe in the autumn, due to the

Wild herbs can be every bit as tasty as those grown in the garden. (Courtesy: Philip Watts)

fact that squirrels and other small mammals have generally been there before you. Walnuts are found in some places as are chestnuts. The possible uses for blackberries are endless: to make jellies, chutneys, add to sauces or to fill a Kilner jar with fruit, top it up with white wine vinegar and leave in a sunny, warm position for a couple of weeks. Strain out the fruit so that you are left with a pure liquid before bottling and storing it in a cool place to use as a dressing over salads. Try experimenting with juniper berries, bilberries and, if you can find them, cranberries (found wild in certain areas). Crab apples and orchard fruits also make a good basis for jellies and chutneys that can be served with many dishes. Elderberries can also be used in a number of ways: boiled in spiced vinegar, they make a superb relish but, of course, their best-known use is in the making of wine.

A glut of fruit can be made into jams and preserves, or even liqueurs and wines.

Liqueurs, Wines and Punches

Country wine is made from the flowers, berries, leaves and roots of plants that grow wild. The purest and most perfect of them are easily made and with the simplest and cheapest of equipment. To make elderflower wine, for example, you would need only florets (how many depends on the strength of flavour you like in your wine), boiling water, dried yeast, a quantity of sugar and assorted fermentation jars, bottles, a funnel, some tubing with which to siphon off the wine, a plastic bucket with a lid and the patience to wait a year while it matures. The only limit to making a

good liqueur from what is available in the countryside is your imagination. Almost anything can be used, but the most popular drinks are those made from sloes (the fruits of the blackthorn), wild damsons, cherries and blackberries. The spirit is usually gin or brandy, although vodka and those lethal spirits that you once brought back from holiday and now sit lurking at the back of the drinks cabinet can also be used to good effect.

The simplest method is to place fruit, sugar and alcohol into a Kilner jar or some similar container with an airtight lid. Place it on a windowsill and shake it daily for about a month before straining, bottling and storing it in a cool place for a few months more. Extra flavours may be gained by using honey instead of sugar or adding herbs such as marjoram and spices such as nutmeg, ginger

Free-range eggs are much preferred by the customer and will sell quite easily. (Courtesy: Rupert Stephenson)

The making of preserves is a good way of using up a glut of garden produce or of successfully utilizing a harvest from the hedgerow. (Courtesy: Philip Watts)

or cinnamon – use chunks or grated, rather than powder, as it is impossible to strain out the powder, which will add an unpleasant, sawdusty taste and feel when it is drunk. Sloe gin made in early October should be ready to drink by Christmas, but the longer you keep it, the better it will be.

'A BIT ON THE SIDE'

No, we are not thinking of an unhealthy and probably dangerous liaison with the partner of a fellow smallholder, more of the fact that, provided that you are not contravening any government or local legislation by doing so, there is the possibility of a little money to be made by selling off surplus produce. Making what used to be known as 'pin money' by the selling of eggs (traditionally the perquisite of the farmer's wife) and growing for the salad

market have already been discussed, but there are other less obvious ways of earning a little cash from even the smallest of small-holdings.

Jam-making usually begins as a result of a glut of fruit. Rather than waste it, it makes sense to turn it into jams and preserves, either for future, personal use or to sell at the local smallholders' meetings. It also makes an acceptable gift at Christmas time. When the garden surplus runs out, it should be possible to make traditional hedgerow jams from blackberries, damsons and suchlike.

A friend of one of the authors used to help his neighbour with a market stall and, in return, the neighbour used to allocate space on a stall for the friend's flowers. One spring he had little for sale, or rather not enough of one kind to make up bunches. His enterprising wife made up some small, mixed posies of odds and ends which sold well. and so he continued with the practice throughout the year using whatever was in season, augmented with a piece or two of old man's beard or wild rose hips taken from the hedges surrounding his fields. No matter what the contents of the posies, the price was always the same and the couple quickly built up a regular clientele who just wanted a posy of seasonal flowers rather than the more expensive bunches of single varieties.

If you are lucky enough to own woodland not only can you periodically coppice hazel for pea sticks and bean poles, but it is worthwhile to consider the possibility of producing some garden furniture from the same source. You may well need logs for your own fire, but, even if you do not, there is a ready market in supplying logs to others, many of whom like the idea of a burning fire even if their home is normally centrally heated by oil or gas.

SHOWING AND EXHIBITING

One day, when you grow that perfectly straight carrot, produce a batch of jam far better than any which can be bought from commercial sources, look into a pen of almost mature poultry (only to realize that one of them is, in fact, a swan rather than an ugly ducking), or glow with a certain pride when a fellow enthusiast who has been at it much longer than you, remarks rather wistfully, 'That's not a bad calf/lamb/kid you've got there', you will be bitten by the showing bug.

In a contradiction of terms, there are, of course, professional 'amateurs', who grow and produce purely for the exhibition side of their hobby – and there is nothing wrong with that. To compete at that level, however, you need to adhere strictly to standards and, to our minds at least, that takes away from the ethos of smallholding. Growing vegetables and producing healthy stock is what it should be

Another way of making money might be from the manufacture of rustic garden furniture.

Showing and competing of any kind should, above all else, be fun.

all about, and if, on the way, some prize-winning features appear, then that is the level at which one should compete. It is, after all, supposed to be fun: an adjunct to the main purpose, but it is an aspect of smallholding that, nevertheless, has a quite subtle bite.

Fascinating though large national competitions can be, it is far better to start small and work up. There is much to learn in doing so and a great deal more fun and enjoyment to be had on occasions where competition is not taken quite so seriously. Obviously, the exact procedures on how to enter vary and depend on what you are exhibiting. Generally, however, you will have applied for a schedule from the show secretary and returned the entry form several weeks beforehand. There are exceptions where this is not necessary – a small, local flower and vegetable show may permit entries on the day, but in the main,

pre-entry is an essential requirement. Always check that you have read the schedule carefully and are entering, not only in the right class, but are also adhering to the criteria given.

The Flower and Vegetable Show

Can there be anything more British than the local flower and vegetable show? Held traditionally, under canvas on a sublime summer's afternoon, the marquee is full of the heady aroma of scented blooms and the tables groan with blooms, flower arrangements and scrupulously scrubbed vegetables. Delving deeper, one finds auxiliary classes: ones which cater for jams, cakes, home-spun woollen products, decorated eggs and country-orientated photographs – something for everyone in fact.

That is the popular image, and, excellent though it is, the fact remains that such shows

Auxiliary classes cater for everything from jams to decorated eggs, or even eggs used as decoration. (Courtesy: Rupert Stephenson)

are also held in the dreariest of municipal halls and community centres (but are, of course, no worse for being so). Organized by local horticultural clubs, they often take place in the spring and the autumn, the produce on show being dictated by the season. Whatever the produce, be it vegetables or flowers, the things to watch for are condition, uniformity and colour; the produce should be free from blemishes, pests and diseases. Symmetry is prized – you will find no knobbly vegetables. There is a great deal of artifice in the world of vegetable showing, the large onions are tied at the top, smaller ones are laid out on a round, sandy bed, like a game of solitaire, carrots are shown in sets of four, with green shoots trimmed like flat-top haircuts, and with long extensions called whips – an effect gained by growing the carrots in long barrels through coarse sand to elongate the tap root.

It is not necessary to be organic on the showing circuit: the attitude of many growers'

pro-pesticide views might disappoint those hoping for self-reliance, but there is, somehow, something nevertheless pleasing about the perfect, symmetrical vegetables they produce: cauliflowers like cumulo-nimbus clouds, mega-leeks with bulbs the size of babies' heads. Bringing you back down to earth, leeks have to show a clean 'V' where the green leaves meet the white tap and even the volume of the bulb is taken into account. With so much to consider, it almost makes you wonder whether showing is worth it … but it is.

Livestock Shows

Here much depends on the type of livestock you are intending to show, the size of the show and the type of class you are considering entering. Exhibiting poultry is a relatively simple matter, but, even so, you must several weeks beforehand begin conditioning your stock, perhaps with the aid of a few

Provided that you have the right quality of stock, the exhibiting of any kind of poultry is relatively easy. (Courtesy: Rupert Stephenson)

special tit-bits, such as linseed, in their diet and also training them to become accustomed to an exhibition pen. Some judicious handling will help a bird to accept the alien conditions and encourage it to show itself in a style demanded by the breed standards. Poultry are also easy to handle – you catch them up, put them in a suitable traveling box, disgorge them carefully into a pen allocated to you by the show steward and leave the rest up to the judge.

But nothing is so simple for the pig show-man: first, of all he or she needs a suitable trailer and, although stock will be penned at the show, when it comes to exhibiting in the actual class, there is nothing for it but to direct the animal round the ring with 'pig boards' – usually advertising a local food product. At least the exhibitor of cattle, sheep or goats fares slightly better in that he has some control over his beast by means of a halter and rope. To have any hope of winning, your livestock will, or certainly should, have had weeks of preparation, not the least of which is getting a particular animal accustomed to being led and showing itself well in public.

Other shows and competitions have their own complications. A milking competition, for example, may necessitate your having to be present the evening before the show. This

obviously entails some forward thinking on your part, one aspect of which is the need to have someone reliable left in charge of your non-show animals at home. On top of this, you will need to take with you buckets for food and water and one for milking into, hay nets, hay, concentrates, fresh foliage and/or grass. For yourself, you will need an overnight bag and a sleeping bag, plus an aptitude for being able to get a night's sleep even though you will be sleeping on site and most probably on hard ground.

'Tricks of the Trade'

Feeding is important and no amount of grooming will disguise the fact that an under-fed animal will not have the sleek and shiny appearance of a specimen that has been care-fully prepared nutritionally for the event. A few of the old-timer's tricks may here come into play: goats, for example, can be made to stand well by the way in which their hooves are trimmed and really good handling can make a lesser animal of any type look good. There are so many of these tricks that it is impossible to mention them all here, but the fact that they exist is another good reason to get to know some experienced and friendly members of your local club or society. However, do not blame any who, understand-ably, deliberately hold back some important

Exhibiting any large livestock involves a great deal of preparation, but is well worth the effort.

detail, not wishing it to be widely broadcast. Also beware of the fact that some of these tricks of the trade may not be strictly legal, there is a world of difference between trying to enhance and trying to deceive. It is not unheard of in the world of flower and vegetable showing for certain unscrupulous individuals to insert soap into slug holes in potatoes or to insert florist's wire inside a drooping daffodil head or for the poultry fancier to add dyes to the washing water, stain the legs with iodine or mask a white spot on the lobe with a touch of lipstick – all of which are 'illegal' practices. On the other hand, it is perfectly acceptable to wash potatoes in milk to give them a richer cream colour, dust onions with French chalk or talcum powder to get a better finish or to wipe down the feathers of a poultry exhibit with a silk handkerchief in order to impart a lovely silky sheen.

HOLIDAYS

No matter how keen you are to embark upon the creation of a smallholding, there will come a time when, either because of a family problem or merely the desire for a change of scenery, you will need to leave your plants and livestock in the hands of others. With any kind of livestock it is never easy to get away, and particularly for the smallholder with a young family who is forced to take his break during the school summer holidays when plants are beginning to ripen and there is likely to be immature stock about. So, what can be done? If you have kind neighbours with an interest in your hobby, then, with the promise that they are welcome to the ripening crops or eggs from the poultry run or that you will look after their greenhouse/budgie/pet mountain lion, or whatever when they next go away, it is perhaps not too much of an

151

Most types of chicken or bantam require a bath in warm water and soapflakes before being exhibited, a practice that is an accepted trick of the trade. (Courtesy: Rupert Stephenson)

imposition to ask them to feed and generally keep an eye on your stock while you are absent.

Lists and Explanations

You should, of course, spend a fair amount of time explaining your routines and ensuring that your helpers understand what is being asked of them. Make the job as simple as possible for those who are keen and enthusiastic but have no personal experience of smallholding. As well as explaining verbally, both the morning and the evening routine, it does no harm at all to leave a typed list of the jobs to be done and then pin it somewhere in the feed shed or, if the neighbours have access to the house, on the kitchen table. Make sure

that the list contains the telephone number of a fellow smallholder somewhere in the area, it is far easier for anyone concerned about a particular problem to ring and ask someone to come and give his opinion and help within an hour or two than it is for the neighbour to ring you. Yes, you may be able to reassure them over the telephone, but otherwise what good can you be a couple of hundred, or even thousand miles away? It should go without saying that the list should also contain the vet's telephone number and, if you are of a particularly efficient nature, you could even ring your vet before you go away in order to inform him that someone will be looking after your stock for the next two weeks – this should make explanations easier in the unhappy event that that someone does, indeed, have to call the surgery when you are away. It is, of course, imperative that you ask your fellow smallholder's permission to leave his telephone number on your list.

You may even find it worthwhile to colour-code your feed bins, or at least identify the contents of each by attaching a label or by writing on the lid with paint or an indelible felt pen marker. We know of several poultry and livestock owners who use plastic bins kept inside the shed or outhouse that have been bought in different colours specifically to aid instant identification. They claim that not only does it make it easier for their neighbours when they are asked to look after stock, but it is also useful to the owners themselves when, on a Sunday morning, they venture out to feed after a heavy night the evening before.

If your smallholding is isolated, it may be difficult to find willing neighbours to keep an eye on things in your absence, in which case, there will be no alternative but to procure the services of a professional team of house-sitters.

Making Things Easy

If you happen to live down a lonely track out in the country and have no immediate neighbours, finding someone to take on the job while you are away may not be so simple. Perhaps a branch of the family, relations or very good friends may be prepared to come and house-sit for the period? Failing that, is it worth considering employing the services of a professional firm of house-sitters? It will, of course, be an expensive addition to the cost of your holiday, but at least your plants and animals will be looked after properly (most of the people who undertake these duties are experienced in all manner of livestock management), and there is the added reassurance that your home will be more secure in your absence.

What other things can be done to make life easier for your locum while you are out sunning yourself on some exotic beach? You could consider setting up an automatic watering system, which might go some way towards easing the burden not only for anyone looking after the vegetable plot or green-

house, but also for yourself throughout the year. There are several types to choose from, but whichever you eventually decide on, make sure that you strip it down periodically, cleaning any valves and flushing through the hoses. It will also pay to keep a stock of spare valves.

For the chicken run there are also commercially produced, automatic pop-holes available. They react to light or are set on a timer and seem at first to be a good idea – maybe for the odd evening when you know you are going to be out late, – but it is probably not wise to trust one over a period of time, and, if you do install such a device, you will still need someone to check that the system has not failed and that the birds have not been locked in all day or, worse still, locked out at night.

With larger livestock, it pays to keep everything in a state of good repair: a gate that falls off its hinges and traps your ankle every time you go through it might not bother you, but you can be sure that outside helpers will be a little less enthusiastic the next time you call on their assistance. Larger animals such as donkeys, sheep and goats might well be beautifully behaved for you, but they will, like children, certainly push (sometimes literally) to see what they can get away with when being looked after by a stranger. A gate that will not close quickly and efficiently in order to prevent an escape could, as a consequence, result in stock getting on to the public highway – a potentially fatal situation to be avoided at all costs.

In the interests of accessibility and maximum security, gates and doors should be kept in a good state of repair.

Glossary

To include here all the terms one is likely to encounter in smallholding would be to take up the whole book. Here are some that may have been mentioned without explanation in the text and will undoubtedly be a part of other books on the various subjects and certainly when talking with like-minded enthusiasts.

AI   artificial insemination, most commonly used in cattle breeding

annual   plant type that germinates, flowers, seeds and dies within the space of a year

ark   small, movable home for poultry or pigs

baconer   a pig of nine months or older

biennial   plants which grow in the first year and seed in the second

breed   any group of livestock whose characteristics distinguish it from any other and which are transferred to subsequent generations

breeding pair/trio   should be a male and the appropriate number of females

brooder   artificial heater for rearing young birds

brood patch   bare patch of skin on the breast of a bird sitting on a clutch of eggs, which, being well supplied with blood vessels, acts as a hot water bottle for the fertile eggs

buckling   year-old, 'entire' billy goat

bulling   behaviour of a cow when ready to be serviced by the bull

cappings   beeswax taken from the top of honey cells each time honey is taken

catch crop   fast maturing vegetable, such as radish, lettuce or early carrots grown in ground that would otherwise stand empty for a few weeks between the harvesting of one crop and the sowing of another

coop   small hutch usually used to house a broody hen when sitting and/or her chicks

coppice   small wood which is cut periodically

cordon   way of training the branches of fruit trees in order that subsequent fruit will be easier to harvest

debeaking   trimming back a bird's upper mandible to prevent feather or vent-pecking

drawing   to clean out the intestines from poultry

dropping board   removable board fixed under perches to catch excrement

entire   male animal that has not been castrated

espalier   *see* cordon

ewe   mature female sheep, usually left as breeding stock

finishing   serious livestock breeders often have a regime for 'finishing off' their stock before slaughter, it almost always involves additional, high-protein feed

flank   side of an animal from the ribs to the thigh

fly strike   infection of maggots laid by flies, which usually affects the rear of a sheep

force moult   artificially persuading a bird to moult at an unnatural point of its cycle, sometimes carried out to ensure that a bird is in prime condition for a show on a given date

frost pocket low-lying area of ground where frost accumulates

gizzard stomach of birds in which the food is ground

hanging by hanging meat after it is killed, the meat is tenderized through changes in the molecular structure resulting in a more tender joint

harden-off means of gradually exposing plants to the harsher elements of outdoors

heavy breed describes poultry breeds whose ancestry possibly derives from the Cochins and Brahmas; despite the name, the Light Sussex is a heavy breed

heel in young trees and shrubs are sometimes temporarily 'heeled in' to a plot of ground before being planted in their proper place; also, the last leeks and parsnips in spring are often moved and heeled in to allow cultivation of their bed

hive tool blunt, broad-ended chisel used by bee-keepers for prising combs apart

hogget yearling sheep of either sex

hybrid in poultry, result of a cross between two or more breeds (such a bird will not breed true and reproduce chicks in its own likeness); in gardening, a plant which is the result of fertilization between two different parents (cross-fertilization)

in-breeding when members of the same family are bred from through several generations; experienced breeders occasionally practise it in order to re-establish a good point in a bird's or animal's make-up

kindling term used by fanciers to describe the birth of rabbits

ley can be permanent or a short cycle, but either way, a description used to describe grass mixtures grown for grazing or hay-making

light breed used in poultry terms to describe breeds which are usually Mediterranean in origin, where their light bones and quick feathering makes them adaptable to hot weather; generally good layers but can be somewhat flighty

line-breeding an understanding of basic genetics is useful before undertaking line breeding, literally within the family line but without in-breeding

moult period when poultry shed their old feathers and grow new ones, generally occurs in the late summer/early autumn

out-breeding opposite of in-breeding; basically, mating different lines of the same breed

perennial plants which grow from year to year

picket tether used for tethering a goat by collar and chain to a stake driven into the ground; this method is becoming less popular due to the limitations of grazing and the potential dangers to the animal; most goat-keepers prefer to keep their stock free-range

poaching grazing livestock out on wet ground for any length of time causes poaching, particularly in gateway areas and around feeding troughs

pollard tree which has been cut off at about 6ft 6in (2m) from the ground

porkers pigs killed at four or five months of age

primary feathers long, stiff feathers at the outer tip of a poultry wing (there should be ten of them)

propolis resinous glue collected by bees

ram mature, entire male sheep, kept for the sole purpose of breeding

rhizome swollen underground storage root seen in several forms of grasses

root crops turnips, swedes and the like, which are often grown as winter feed for livestock

rotation to grow crops in different parts of the garden or smallholding in order that the different varieties impart their own particular benefits to the soil (peas, for example, are rich in nitrogen). rotation also minimizes the risk of disease

secondary feathers quill feathers on the wing, which are usually visible when the wings are either folded or extended

served when a female stock animal has been inseminated she is known as having been 'served'

skep old-fashioned, alternative name for the small straw beehives much favoured by apiarists in the past

spur pointed, horny projection at the base and rear of a cockerel's legs; small nodules are sometimes noticed on the female

smoke gun device with a combustion chamber in which cardboard or sacking are burnt, blowing smoke into the entrance of a beehive will make bees drowsy and less inclined to sting

staring description of an animal's coat with the outer guard hairs sticking outwards rather than lying sleekly along the body, and being 'spiky' in appearance

tedding turning and lifting hay so that it dries in the sun and wind

ticks insects that feed on the blood of animals (including humans); as they are normally found in bracken areas, they are of particular concern to sheep-keepers in these regions, but all country dwellers should be aware of their existence, checking domestic animals on a regular basis

trembles also known as 'shakes', trembles affect piglets from birth, this can be an inherited trait or caused by various virus infections

tupping when the ram is turned in with the ewes in order to service them this is known as tupping

vermiculite treated mineral in the form of translucent flakes, sterile and water retentive, therefore very useful for adding to compost or for covering newly-sown seeds

weaners piglets of seven to eight weeks that have been 'weaned' from their mother

wether castrated young adult male sheep being reared for meat

wing clipping sometimes it is necessary to clip the primary and secondary feathers of one wing to prevent lighter breeds of poultry from flying; the feathers will regrow in the next moult

Useful Contacts and Publications

All these contact details were correct as at summer 2008, but telephone numbers especially are obviously subject to change. Note that the secretaries of some organizations are volunteers, so treat them with kindness, remember that they probably have families and choose the least likely inconvenient time to make contact.

Almost every county seems to have its own smallholders' association – there are far too many to include here. Type 'smallholders' associations' into an internet search engine and you should find what you are looking for, but, failing that, contact your local library who may be able to help.

British Beekeepers Association: www.bbka.org.uk
British Cattle Movement Service: 0845 050 1234
British Goat Society: 01626 833168; www.allgoats.com
British Pig Association: 01923 695295; www.britishpigs.org
British Rabbit Council: 01636 676042; www.thebrc.org
Brogdale Horticultural Trust (home of the National Fruit Collection): 01795 535286; www.brogdale.org.uk
DEFRA: 08459 335577; www.defra.gov.uk
Codes of Recommendations:
www.defra.gov.uk/animal/welfare/farmed/on-farm.html/legislation
Domestic Fowl Trust: 01386 833083.
Food Standards Agency: 020 7276 8181; www.food.gov.uk
Lantra Connect: 0345 078007; www.lantra.co.uk
Local Animal Health Offices: www.defra.gov.uk/corporate/contacts/ahdo.html
National Sheep Association: 01684 892661
Poultry Club of Great Britain: 01476 550067; www.poultryclub.org
Rare Breeds Survival Trust: 02476 696551; www.rbst.org.uk
Royal Horticultural Society: 0845 260 9000; www.rhs.org.uk
Soil Association: 0117 314 5000; www.soilassociation.org

Periodicals

The following publications are recommended:

Country Smallholding (editor: Diane Cowgill): Fair Oak Close, Exeter Airport Business Park, Clyst Honiton, Exeter EX5 2UL; 01392 888481; www.countrysmallholding.com
Smallholder (editor: Liz Wright): Hook House, Hook Road, Wimblington, March, Cambridgeshire PE15 0QL; 01354 741538; www.smallholder.co.uk

Index